How Life Learned To Live

How Life Learned to Live

Adaptation in Nature

by Helmut Tributsch

translated by Miriam Varon

The MIT Press
Cambridge, Massachusetts
London, England

First published as *Wie das Leben leben lernte* (© 1976 Deutsche Verlags-Anstalt GmbH, Stuttgart).

This book was set in Baskerville
by The MIT Press Computergraphics Department
and printed and bound by Halliday Lithograph
in the United States of America.

Library of Congress Cataloging in Publication Data

Tributsch, Helmut, 1943–
　How life learned to live.

　Translation of: Wie das Leben leben lernte.
　Includes bibliography and index.
　1. Biophysics. 2. Adaptation (Biology)
I. Title.
QH505.T6813　　1982　　574.19′1　　82–10018
ISBN 0–262–20045–7　　　　　　　AACR2

Contents

Preface

This is more than an illustrated book on biophysics for the layman. It is also the story of an adventure that started in the laboratories of the University of California at Berkeley, where I began to ask myself whether today's naturalists, buried under instruments and flooded with scientific literature, have not strayed too far from nature.

My work at Berkeley dealt with the transformation of sunlight into chemical energy by photosynthesis in plants. My studies were conducted not in a sunlit garden, but behind a mountain of electronic and optical instruments. For the most part, my mind had to dwell in a world of abstract models and formulas. Of course that was necessary at such a high level of research, but every time I walked through the woods of the Sierra Nevada and watched my scientific project "breathe," I felt how essential and productive the contact with nature was for me. Progress in research is not based on thinking alone; intuition, inspiration, and chance are equally important. How to put oneself into the way of the most interesting inspiration may be a matter of opinion. I developed my own ideas on the subject.

When I decided to write a book about the application of the laws of physics in nature so as to learn more about the subject, I was convinced from the start that I would have to combine this work with actual experience in nature. I had no desire to merely contribute a mass of information to libraries; rather, I meant to see for myself, and perhaps take pictures of, those processes that interested me.

When my contract with the university expired in 1971, I packed my photographic gear and headed south along the Pan American Highway, the dream road of the Americas. It was to be more than two years before the South American continent would let me go again. During this time I crossed all of its fascinating wilderness areas: the Galapagos Islands off the coast of Ecuador, the Andes all the way to Patagonia and Tierra del Fuego, the coast of Peru and Chile with its bird islands, and the wilderness region of the Chaco in Paraguay and Bolivia. I visited the Matto Grosso territory in Brazil and made my way through the jungle along the freshly cut route of the Trans-Amazon Highway. Finally, I embarked upon a journey of many legs along the entire length of the Amazon River, from its delta to the Andes. At all times, I was looking for interesting biophysical phenomena.

My financial resources were rather limited. I managed to stay within my means by leading the simple life and using the native modes of transportation. For a time I worked as scientific advisor to the mining industry in the Atacama Desert of Chile. I covered thousands of miles by truck or on the deck of some small freighter. Where there were no roads, I traveled for weeks on foot or on horseback. It was sometimes difficult to reach a remote and uninhabited area without shelling out a lot of money to hire a four-wheel-drive, a boat, or a plane. Often I had to wait a long time for an opportunity to share a ride, and quite often I had to take a chance and ask to be dropped somewhere in the wilds in a spot that seemed interesting. Some of the pictures in this book were taken under rather hazardous conditions; but my adventures were well worth the trouble, and they made dealing with scientific problems fascinating and productive.

This venture, these experiences with man and nature, changed my entire attitude toward the natural sciences. I now see them as much more than a testing ground for a handful of competing scientists. First and foremost, they ought to be a source of enjoyment and inspiration for the many who are interested in nature.

Also, biophysics could become a pioneer of modern science, and not only in the transformation of visible light. With the transpiration system of the trees nature has succeeded in converting the heat of the sun into mechanical energy. Mangrove plants use their suction and transpiration apparatus to desalinate sea water. Why should we not try to duplicate these mechanisms with our technology?

Mankind invests enormous financial resources and the most sophisticated technologies to guard against the perils of atomic energy, while all around us enormous quantities of clean solar energy are going to waste. Is this venture worth the price? The expansion of atomic-energy supplies has become a matter of controversy. Energy from the atom is more expensive than had been expected, represents a potential source of danger, and creates radioactive waste that will be a burden to future generations. Moreover, atomic energy infringes upon the secret rule of nature's energy household. Energy should be inexhaustible, clean, and convertible in the smallest units.

Only solar energy complies with these conditions. In tropical zones and deserts an average of some 250 watts in luminous energy reaches each square meter of soil. In central Europe the figure is still above 100 watts. The radiation energy of each square kilometer could, if harnessed, supply a respectable power plant. Green plants tap no more than 2 percent of irradiating solar energy, yet they are adequately provided. Why does man not turn to solar energy? Why does he not try to emulate the plants and build chemical energy from the sun? There is an abundance of infertile land and unused expanses of water available which could serve as windows for power plants. No scientific reason can be proffered against the possibility of photochemical energy conversion, but we do find explanations why so far only modest progress has been made in the area of solar energy conversion. The problems are too difficult to be tackled by individual scientists and small groups, and immediate success is too improbable. And who likes to sacrifice his own career to others? For a breakthrough, strong financial and psychological motivation is needed that would sweep along many capable scientists and large institutions. Atomic energy got such a push because of war and military confrontation. Perhaps the present energy shortage will smooth the way for economical use of solar energy. Let us hope that we won't be too late in remembering that nature has been operating a similar process for millions of years.

People in general think of nature and physical technology as opposite and conflicting worlds. Many live only in the technical, others only in the natural world. Both groups advocate their own interests and convictions. Why should it not be possible to work toward technological progress that is in harmony with nature? The schizophrenia from which we suffer is rooted in our schooling. Our biology studies disregard the fascinating technology of nature. Our studies in physics or technology pay just as little heed to the ingenious technical inventions in the animal and plant kingdoms. Would not an engineer familiar with the technological ideas of nature have more respect for nature? Would not a naturalist who realized that life requires technology have more understanding for technological progress? Knowledge that reconciles nature and physical technology could pave the way for the progress of technology not as a threat to man, but to his benefit.

How Life Learned To Live

1

Technology Against a Life-Threatening Environment

Life came into this world more than 3 billion years ago. It began somewhere as a helpless combination of organic molecules in a very harsh physical environment. Everything seemed to conspire against survival: the strong solar radiation, which could split the delicate chains of the molecules; the cold, which could freeze life's first cautious movements; the heat, which could turn carefully ordered structures back into chaos; and the pressures, which could easily squeeze fragile cells to death.

Inorganic nature had stretched a deadly net of unalterable physical laws around life's delicate beginnings. Life's only chance to prevail lay in learning to master these forces. The problems it had to face were overwhelming. For one thing, in order to develop it had to tap the only inexhaustible source of energy: solar radiation. But how could particles of light be caught, and how could their energy be stored for use as needed? Primitive life never asked these questions. It possessed neither creative intelligence nor the will to solve problems. Propagation was the sole driving power behind the evolution of life.

Life drew its creativity from a stock of chance mutations in the genetic code of successive generations. Those descendants that were better able to withstand the environment survived and passed on the technical knowhow by which they had slipped through the loopholes of physical laws. They developed methods and mechanisms by which they were able to cope with the hard struggle for survival. Competing for food and space, the living creatures began to spread and to colonize the most hostile regions of the Earth: the dry, heat-baked deserts, the oxygen-poor highlands, the atmosphere, the frozen arctic and antarctic terrains, and the darkness of the oceans.

How has life fared in its confrontation with physics, which started so unequally? To what degree has life succeeded in exploring the laws of inanimate nature and using them to its advantage?

A turkey buzzard above the Peruvian desert. Life sends its front-line patrols into the most inhospitable regions of the Earth.

Lichen communities of algae and fungi have come to within 400 kilometers of the South Pole. There they exist at an average temperature of −15°C. In the Himalayas, they are found at an altitude of 7,000 meters (not high enough, though, for an altitude record; migrating geese fly over Mount Everest at 8,800 meters). Numerous large and small living creatures of the subpolar regions of North America and Asia survive temperature lows of −50°C. On the other end of the scale, algae live in hot springs at 85°C and fish at 50°C. Life has also conquered the world of high pressures. Whales dive to depths of 1,000 meters, where each square centimeter carries the weight of 100 kilograms. They have conquered decompression sickness and are able to remain underwater for 1–2 hours without renewing their air supply. The swiftest land creatures, the cheetah and the ostrich, reach speeds of 100 kilometers per hour, and some hawks fly at 200 kilometers per hour. Specialized sea predators such as the tuna swim at speeds of over 40 knots, or about 75 kilometers per hour, and can easily overtake atomic submarines. Small golden plovers (sandpipers) from Alaska and Canada fly nonstop 3,500 kilometers across the Pacific to Hawaii or across the Atlantic to South America for a winter vacation each year. The arctic tern, steering by mysterious navigational signals, covers up to 29,000 kilometers on its annual round trip between the

Arctic and Antarctica. Some insects move their wings up to a thousand times per second. In the darkness of the night, bats hunt for insects with almost infallible precision thanks to their ultrasonic equipment. Some snakes locate their prey by means of infrared-sensitive organs. There are fishes that use electrical fields to scan murky river waters.

As these examples show, life has met the physical challenge with flying colors. But how were increasingly complicated biophysical systems able to develop at all in the course of evolution? A fundamental law of thermodynamics is that irreversible processes—such as life—occur only when the entropy (the degree of disorder) increases thereby. Heat distributed evenly through space no longer flows back to its source, and an escaped gas never condenses by itself. Life does not break any physical laws, but it makes very clever use of ways to circumvent them. The natural law of increasing entropy applies only to closed systems (systems that do not exchange energy with their environment). Living organisms can develop toward increasing order and complexity because they absorb energy from outside. They organize themselves at the expense of increasing entropy in their surroundings. As soon as a living creature ceases to absorb energy, it must submit to the law of entropy and disintegrate: It dies.

It is entirely realistic to assume that human technological and scientific progress is polarized. Why, for example, do we build atomic rather than solar power stations? From the scientific viewpoint, there is nothing to be said against using solar energy to produce fuels. Nature provides a model in the photosynthetic mechanism of the plants on which all higher forms of life depend. If nonetheless an enormous investment is being made in fission, the reasons are to be found in history and psychology rather than in science. Perhaps nothing more than an encouraging chance discovery or some propitious political situation would have steered technology in the other direction. The "state of the art" merely indicates the actual development on the basis of some chance conditions, and says nothing about potential alternate achievements that could have been arrived at with the same effort. To try to imagine what other roads technology could have taken may be a difficult and speculative endeavour, but nature could serve as a guide. The statistical patterns according to which nature's techniques develop allow for a much more objective unfolding of all potentials. They are limited—by natural selection—only insofar as they are guided by the advantages they bring to life. Such a selection should be equally favorable to man.

This book will show, by means of noteworthy examples, how life has learned to master the physical problems posed by nature, and how it has chosen ways that sometimes differ widely from those of man.

2

Mechanics and Building Techniques

Architecture and materials

Three guidelines of modern architecture are equally valid for biological forms: A building should be the realization of clear, elegant structural ideas; it should be solid and economic; and it should be esthetically harmonious with its environment. It is difficult to give intellectual reasons why we regard many forms of nature as esthetically pleasing. The fact is that biological structures have been subject to ruthless selection through millions of years and have been custom-designed with great precision for their specific purpose. One would need only to make the proper choice among the various forms of nature in order to solve some architectural problems without painstaking technical research.

A famous example is the water lily *Victoria amazonica*. The supports and reinforcements on the undersurface of its wide floating leaves served as a model when Sir Joseph Paxton designed the Crystal Palace of London in the middle of the nineteenth century. It is hardly probable that biology also inspired Fuller's self-supporting dome, but it is interesting to note that the dome's assembly of triangular or hexagonal elements is found among the siliceous structures of tiny diatoms. Honeycomblike shapes and fluted structures, too, are frequently encountered in nature. A vast array of models is available in nature for sensible ways of packaging, for mechanical strutting, and for geometrically optimal layout.

Still, some of the architectonic tasks nature has to face differ considerably from the aims of human architects. Moreover, we are talking about structures whose functions are not altogether understood by biologists. (This is the case with the wondrous diatoms, which—although they are mostly suspended in water and filled with liquid, and thus

The giant *Victoria amazonica* unfolds its true splendor only in its natural environment: the quiet, jungle-shadowed backwaters of the Amazon River.

are subject to very little compression—seem to be perfect models for self-supporting domes.) Besides, most of the fascinating biological structures owe their shape to physiological necessities rather than to those of statics or architecture. For example, some are designed in such a way as to be well able to carry matter, absorb light, reflect heat, and breathe simultaneously.

Except for cases in which it is possible to define unequivocally the physical functions of biological structures, nature is a source of constructive and artistic inspiration rather than true architectural solutions. What architecture ought to learn from nature, above all, is the art of coping with a host of technical problems that require simultaneous optimal solutions. Architecture cannot afford to focus on the construction of sophisticated housing projects and factories without at the same time coming to grips with problems of transportation, pollution, leisure, and social contact.

Though it is not easy to imagine how nature would go about building a metropolis, it is certain that there would be little resemblance to one of our population centers, which are far from being harmoniously functioning organisms. Though human life depends on the plant world as the supplier of oxygen and food, vegetation has to a large extent

The undersurface of a *Victoria amazonica* leaf.

been exiled from the cities. Precious energy from the sun is wasted on roofs, parking lots, and streets, and dust and exhaust gases impair breathing. Nature certainly would make a radical attempt at providing optimal conditions for life.

Since life needs light, air, and a protective shield, it is in theory subject to conditions similar to those that prevail for a photochemical surface reaction. Such a reaction is the process of photosynthesis in green leaves, by which light is transformed into chemical energy. Perhaps, then, nature would build cities similar to the submicroscopic thylakoid structures—the power stations of plants, which consist of self-contained flat membrane sacs, often stacked like rolls of coins and linked to each other by many cross-connections. The units are arranged so as to make maximal use of light and to form as large a contact surface as possible with the environment—architectonic criteria our cities still fail to meet adequately.

A bird's-eye view of a natural metropolis would show nothing but green. No roofs, parking lots, or highways would be visible. All flat

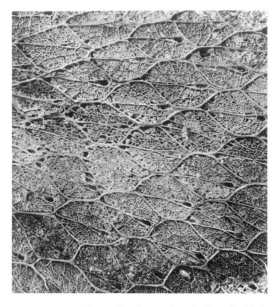

Structures on the stalk of a rotting Indian fig (Galapagos Islands).

surfaces would be covered with woods, parks, and gardens. The vertical structures would be the facades of offices, residential buildings, cafés, and boutiques, all with access to nature. Inside the "thylakoid structures" would be sufficient space for transportation, parking lots, shopping malls, and factories, which could manage with artificial light.

A building material must have strictly defined physical properties once it is put into a structure. Sometimes toughness and brittleness are required, sometimes strength and elasticity, sometimes good heat insulation. Paper, cardboard, and wood fiberboard, which have very good insulating properties, are made of small particles of wood and straw glued together with some bonding agent. Such building materials are used by several large families of animals. Wasps build their featherweight structures with wood, which they gnaw off and then mix with saliva. The reason their nests are usually gray is that they use weathered wood, which is relatively easy to loosen. Numerous species of ants (*Lasius*) build big cardboard nests which are subdivided into complex warrens of corridors and chambers.

The potter wasp (*Eumenes*) uses moist loam to build mud containers for its brood. If the mud is too dry, it is moistened before being scraped off; then it is transported in small pellets and drawn into thin strips

The partially exposed "cardboard" structure of a termite colony (Amazon region, Colombia).

before being used to build the wall of the container. The oven bird also uses moist mud for its nest, but adds plant fibers as reinforcement.

Mortar too has been used by animals. In the region of the Amazon one sometimes sees clay vessels swinging from branches in the middle of the wilderness. These are the nests of a species of wasps (*Polybia*) that mix clay and sand to build their castles. The leaf-cutting bee (*Chalicodoma*) uses dry stone dust or sand and mixes it with saliva, forming small mortar pellets to build its brood cell.

The most efficient builders are the termites. Their structures are as hard as concrete and often several meters high. In tropical areas of Central and South America, termite structures are especially eye-catching wherever settlers have cleared land by burning; among the charred trunks stand the black lumps of termite fortresses. Termites often use bits of earth, sand, or wood as building material, which they glue together with their droppings and their saliva. Some termites (*Apicotermes*) build their castles entirely out of their excrement, which dries quickly and does not decay.

Walls fitted together of natural stone are also to be found in the animal world. The South Asian jawfish (*Gnathypops rosenbergi*), which digs vertical holes, uses stones and shells to keep the walls from caving in. The caddis worm, which lives on brook and pond bottoms, often

"Log cabins" of silk spun by the larvae of the bagworm moth (Paraguayan Chaco).

reinforces its living quarters—funnels woven of silk—with a wall of small pebbles. At times it also builds in bits of reed and wood. In the cocoons of bagworm moth caterpillars (*Psychidae*) we find "log cabins,"— wood splinters assembled with silk thread in various patterns.

Plastic building materials are also evident in the animal world. The swallowlike swiftlet (*Salangane*) of South Asia builds the transparent nests used in Chinese bird's-nest soup with its own saliva, which it applies gradually to a rock base with its tongue. Less welcome in the kitchen are nests that contain moss, plant particles, threads, and feathers. Those, however, are the nests that are found most frequently and that are of special physical interest. They are stronger, since the synthetic is reinforced by the fibers. The Old World palm swift (*Cypsiurus parvus*) likewise uses wood and plant fibers to carefully reinforce its saliva nests, which it glues to the leaves of palms.

Reinforcement by the planned use of fibers, as in fiberglass or ferro-concrete, is also evident in the thin cardboard pillars of wasps' nests and honeycombs. In principle, these pillars consist of the same material as the rest of the structure. However, they derive their great strength from the fact that all the wood fibers are arranged in a parallel pattern. That is to say, the wasps instinctively take into consideration the strength requirements of their building materials while building their nests— and they do so with ingenious simplicity.

Modern industrial packaging makes much use of featherweight synthetics, such as Styropor with its many enclosed air bubbles. Such synthetics are not only very stable, they are also shockproof and heat-insulating. Similar foams have long been in use in nature. In order for foam to be produced, the surface tension of the liquid material must be lowered. Since large surfaces absorb dirt easier, soaps and detergents are also foam-producing materials. The Javanese flying frog produces foam around its eggs during the mating process. For this purpose it excretes a slimy liquid which it beats into foam with its hind legs. Then the foam bubble is pressed against a leaf and hardens superficially. The slippery tadpoles find optimal conditions for development in this moist haven. Many other frogs (and insects, too, such as the froghopper) blow foam bubbles for their broods.

The bee's use of wax, a fatlike compound, is familiar. These skillful insects also work with resin, which they gather from tree buds and use to insulate their buildings or to embalm intruders they have killed.

A single cocoon of the silkworm contains 3,000–4,000 meters of silk (a protein), of which about 1,000 meters can be spooled off the middle layer. Spiders too use silk for their intricate webs. Tree ants (*Oecophylla*) use their own larvae as sources of silk for the weaving together of live leaves into living quarters.

Nature has experimented with practically every building material man is using today except metals.

Traveling in the Amazon delta through narrow jungle-lined channels, one encounters at intervals the simple pile dwellings of native colonists. Almost always one finds close by dozens of the saclike nests of American orioles (*Icteridae*). These sprightly black-and-yellow masters of the building trade do not seek the company of man because they are bored or have designs on his refuse; rather, man keeps their enemies, beasts of prey and reptiles, at a safe distance. In thinly populated regions of the Peruvian wilderness in the upper Amazon I often saw their nests hanging in very exposed spots high above some torrent. Since the woven nests (which sometimes reach a length of 1.5 meters) are not hidden but are visible from afar, hanging from the flexible end of a branch, they very often are the target of hungry aggressors out for a meal. But the weaving birds protect their nests precisely by attaching them to thin elastic branches. The branches will not hold the heavier predators, and the nimble tree snakes have difficulties getting into a dangling nest with a narrow entrance.

To hang their nests confronted these birds with difficult problems regarding structural strength. They had to learn complicated knots

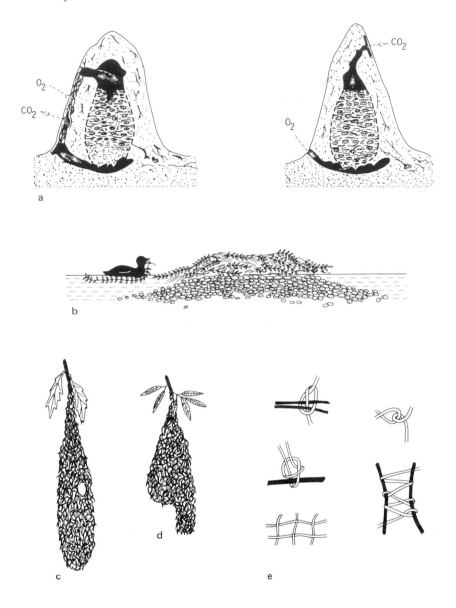

(a) The climate-control system of the Natal termite. (b) The stone island of the horned coot. (c) The nest of the oriole. (d) The nest of the weaver bird (*Ploceina*). (e) The knots used by weaver birds.

Nests of orioles in a jungle tree (Amazon delta).

and weaving techniques with which to manipulate their materials (plant fibers and grasses). Besides the American orioles, there exist ten subfamilies of the true weaver birds (*Ploceidae*) that have developed a rather different kind of beak skill. Then there are some skillful "outsiders," for instance the penduline tit (*Remiz pendulinus*). The true African weavers (*Ploceinae*) are held to be the most accomplished artists, but I was able to see for myself that the less known South American weavers (especially the orioles) are not far behind.

The task of building the woven nest starts with the selection of suitable plant materials. Tear-resistant grasses in particular have the required properties, but even they do not always satisfy the weaver's needs. Often the bird flies to a palm, nibbles at the edge of a leaf, and takes off, dragging a long elastic fiber. This technique seems to be widespread among the true weavers of Africa as well as among the orioles of South America.

In the central wilderness of the Paraguayan Chaco, which is too dry for palms, a farmer and amateur naturalist told me that the weaving orioles extract plant fibers from the slender thorn-tipped leaves of the agave. The leaf margins of this plant are reinforced by a very tear-resistant fiber, which the birds cut and loosen with their beaks. (Later, I saw an old Ajore Indian woman extract the same fiber from agave leaves, to use it for baskets and cords.)

In areas that offer a large assortment of plants, the South American weaver birds use thin tear-resistant air roots, which impart to their hanging nests a pitch-black or pretty reddish-brown color. When weaving, the bird often uses its feet to hold one end of the fiber while it guides the other end with its beak. The method of entwinement varies greatly and is determined by the given technical problem. Loops may be laid around a small branch, or true knots may be tied. Numerous ways of knotting and anchorage used by sailors can be found in the works of weaving birds. Often the fibers are stuck into the web, pulled out again from an adjacent spot until they are used up, and then knotted at the end.

Some of the true weavers, for instance Cassin's weaverbird (*Malimbus cassini*), use a very regular weave; others work rather like basket weavers. They all manage to achieve an astonishing degree of mechanical carrying power in their structures. They could ill afford being mediocre architects: Among many of the true weavers the male alone builds the nest and the female carefully inspects the finished structure before deciding to team up with the builder. Some weavers have been observed building, tearing down, and replacing with better models up to two dozen nests during a single breeding season before they succeeded in wooing a bride. Who knows what heights our architecture might reach if our society were to apply equally rigorous methods of selection!

City planners have known for a long time that a residential center has economic and communal advantages over a multitude of isolated individual homes. Some weaver birds have successfully experimented with community nests. The sociable weavers (*Philetarius socius*) pool their forces to construct a thatched roof in the strong branches of a tree; then to the bottom of this roof they attach their individual living quarters in close proximity. After many years of additional building, the roof may reach a diameter of 5 meters and contain more than a hundred condominiums. The buffalo weaver (*Bubalornis albirostris*) likewise builds huge communal nests. Such structures might easily turn into giant larders for nimble predators, but the buffalo weaver constructs a fortress of thorny branches.

The hanging nest of the penduline tit is more than just a web of smooth fibers. Its close-woven feltlike material is produced by intertwining short soft pieces of wool and down into long supporting fibers in the knotting technique of the oriental carpet weaver. The African penduline tit (*Anthoscopus caroli*) even knots a curtain which it draws in front of the entrance when it leaves its hanging residence. Africans sometimes hang these nests from their belts as purses.

A strong sense of physical-technical correlations is required for the execution of structures that can support the builder and the brood, resist wind, rain, heat, and cold, and give protection from enemies. How was it possible for such complicated instinctive knowledge to develop and be passed on to successive generations?

It is astonishing by what simple means stable, resistant, and pleasant living quarters can be constructed. An impressive example are the adobe houses of South America. At first sight one would say that they consist of nothing but impure clay, but this would overlook the essential reason for their solidity. Those bits of straw and plant particles that can be distinguished in the adobe were added to the moist loam before it was pressed into raw bricks, and they account for the coherence of the loam and prevent the brick from crumbling. Adobe is clay reinforced and stabilized by plant fiber. It is not only the poor *Indios* who have built and still are building their huts with adobe. Highly developed South American cultures have left to posterity richly decorated palaces and temples of adobe. The buildings of Chan-Chan, the capital of the Chimu empire on the north coast of Peru, have for more than half a millenium withstood earthquakes, floods, and unrest without disappearing from the face of the Earth. Adobe churches built 400 years ago by Spanish *Conquistadores* have remained in use to this day.

My interest in the simple elegance of adobe buildings was what drew my attention to the mud nests of the oven bird (*Furnarius*) in the eastern part of South America. These semispherical mud castles high up in a tree or on a pole are a familiar sight in the rural areas of Paraguay and the Matto Grosso. I climbed a likely tree in the Chaco to get a sample. With quite a bit of patience and with the help of a machete I managed to detach the clay base from the branch. There I was, with a mud ball weighing several kilograms in my hands. It was too heavy to be simply tucked under my arm. After a lot of acrobatic contortions, I used my shirt to tie my treasure around my shoulders and descended. When I broke open the hard mud casing and had a look at the many plant particles that had been built in, my suppositions were confirmed: The oven bird has indeed discovered the principle of adobe construction. I was assured by natives that the oven bird purposefully mixes pure clay with grasses or bits of wood when gathering mud pellets. Equally remarkable is the architecture of this potter's fortress. The walls are raised on a mud base in a shape somewhat suggesting a spiral and then domed over. The female sits in the closed inner chamber and is fed by the male through the hole leading to the "lobby."

An oven bird nest of mud and plant fibers (Paraguayan Chaco).

Surely the oven bird preceded man in the art of improving the mechanical properties of clay by adding natural fiber. Could it be that man learned from the bird?

To give some animals a fair chance for survival, nature had to supply a considerable amount of instinctive knowledge of physics. This is particularly true for the beavers, whose habitats are the rivers and brooks of the cool northern woods. These rodents, which feed on bark, foliage, and branches, alter and shape their environment more than any other mammal with the exception of man. Beavers dam up a waterway with logs, thereby creating a pond. Right in the center of this pond they put their lodge, which looks from the outside like a huge heap of twigs and branches sticking out of the water. The sleeping and living quarters are above water, accessible only through an entrance below the water level.

Beavers proceed like engineers with sound knowledge of physics in the construction and maintenance of their installations. Right from the selection of a building site for the dam, they need a great deal of skill so that a large and deep pond can be created with the least expenditure of labor. The beaver cleverly incorporates the environment and natural obstacles like rocks and trees into the construction. Very long dams are built only when there is no way of avoiding them.

One of the longest beaver dams known is located in the Jefferson River near Three Forks, Montana. It is 642 meters long, reaches 3 meters in height, and is 6 meters wide at the base. A rider on horseback can easily travel along this dam. The skill with which the beavers choose their building sites proves that they have a clear concept of how water spreads over rough terrain. They are instinctively aware of the physical law according to which water in communicating containers stands at the identical level. The first phase of dam building requires a remarkable sense for hydromechanics and mechanics. Any child who has tried building a dam knows how difficult it is to do a good job.

Beavers, it seems, proceed according to a technically well-thought-out plan. They first ram individual poles into the bed of the waterway. This part of the work is accomplished not only with strength and effort, but also with a clear concept. The wooden poles are planted at an angle against the current whereby the thick end is anchored in the ground while the thin end ploughs the current. The beaver takes two important physical factors simultaneously into account: the friction in flowing liquids and the mechanical transmission of energy. The thin end of the angled pole offers the least possible resistance to the water, and the thick end can optimally withstand the bending load (which increases toward the bottom). The beaver proceeds with the dam construction by building horizontal logs into the weir of parallel poles directed at an angle against the current. Then the dam is gradually reinforced with twigs, leaves, reeds, mud, and stones. Sometimes forked branches are used to anchor the structure to rocks and trees. Finally, the last sluice of the dam, through which water flows with great force, is blocked without much trouble by crossbeams.

A lot of remarkable detail catches the eye of the physicist. For one thing, a beaver dam does not generally run in a straight line. Rather, it takes the form of two or three contiguous wedges placed against the current. As a consequence, the water does not hit the dam head on; it is diverted and loses force in eddies and currents so that the impact is distributed over a much larger surface of the dam. The same effect is produced by the ditch the beavers dig directly in front of the dam into the river bed; where the water is deep, it flows more slowly and the pressure on the dam is diminished. The beaver even has a routine method to regulate the water level: When the water threatens to flood the lodge, the depth of the dam's spillway is increased; and when the level sinks too low, the dam is improved and made higher.

One more cause for amazement is the instinctive understanding the beaver has for the laws of physical friction. When larger trunks must

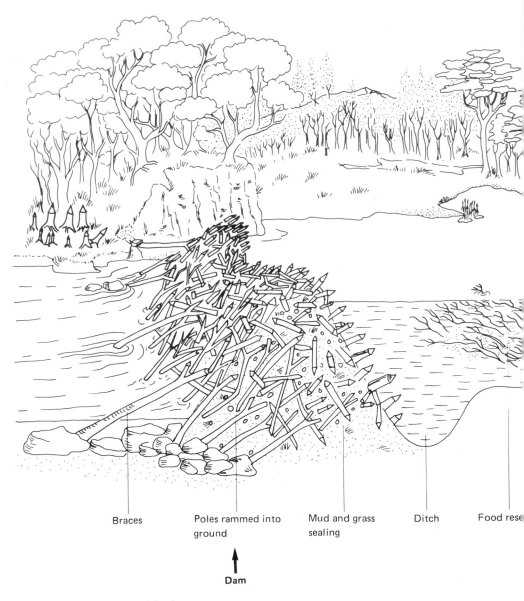

Braces   Poles rammed into   Mud and grass   Ditch   Food rese
              ground   sealing

↑
Dam

Structures erected by beavers.

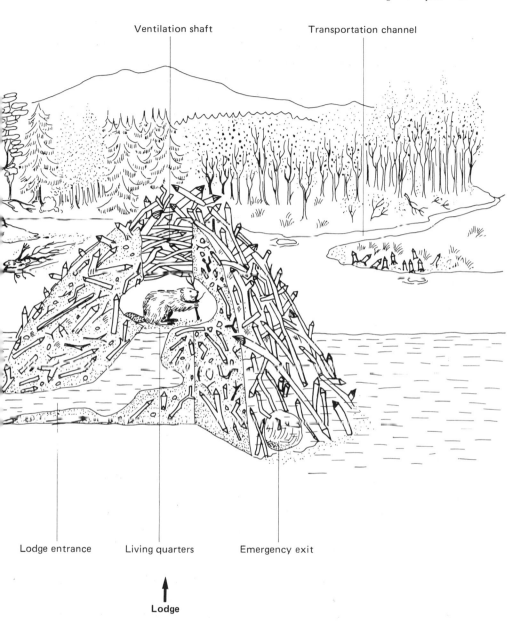

Ventilation shaft

Transportation channel

Lodge entrance

Living quarters

Emergency exit

↑
Lodge

be brought over land to the building site, the path is cleared of obstructions and made smooth in preparation. Where the ground is flat, the beavers often dig channels up to 300 meters long, about half a meter wide, and 30–60 centimeters deep, into which the river water is diverted. Then they raft the trunks over these channels. Sometimes, where the descent is steep, they even feed wells into the channels.

When water cools, its density increases; but below 4°C it gets lighter again, since the short-range order of its molecules changes. Therefore, icing always starts on the surface, while somewhat warmer layers lie below. Beavers, familiar with these properties of water, anchor their winter food supplies of plant matter in the depths of the artificial pond. During the harshest part of winter, when a shield of ice covers the pond, the beavers indulge in a form of recreation that offers proof patent of their mastery over the physical environment: The sluices of the dam are opened slightly and the water level in the pond is lowered. With the ice roof as a cover, the beavers can then cavort in open water.

Beavers' innate knowledge of physics and technology could easily be envied, but we must not overlook its limitations. Their method of felling trees, for instance, is anything but elegant. They chew a circular groove around the trunk of a tree (which may have a diameter of up to 25 centimeters). Then, when the tree starts to crack, they run as fast as they can to the water and jump in. Once in a while, a beaver is killed by a falling tree.

The concretelike fortresses of termites often house several million of those photophobic insects, which leave their homes only at night through subterranean passages. However, the termites and the fungi that constitute their food supply use oxygen continuously. Worker termites do not even have wings to fan themselves, as bees do. How do the termites solve their ventilation problem without windows, doors, or fans?

The *Bellicositermes natalensis* termite of West Africa builds pyramidal nests about 3.5 meters high from which vertical ribs stand out like buttresses on all sides. The ribs resemble the cooling fins of an engine or a radiator, and do indeed serve a similar function. The used-up warm air that rises from the brood chambers and fungus gardens deep inside the fortress is collected on the "floor" above in a sort of air chamber and channeled into the ribs, where it is cooled and from whence it is then returned below through fine passages. The many tiny capillaries in the cooling ducts are separated from the environment only by thin porous walls which allow oxygen to be replenished and carbon dioxide to be expelled. The purified, cooled air collects in a basement room and is then rechanneled into the colony by the air current. During the day, when the sun beats down on the nest, the circulation system (which is based on the temperature-dependent density of the air) is by no means interrupted. The air current simply reverses. Since the air is now being heated in the cooling ribs, it rises; then it descends, while cooling, into the lower portions of the colony. The same termites make do without this complicated system in other regions, where they conduct the used-up warm air out of their nests through porous walls in the upper region and let fresh air come into the basement from outside. It is difficult to imagine how they discovered these laws of air circulation. It is even more difficult to grasp how a teeming army of hundreds of thousands of termite workers can again and again make ingenious use of applied physics.

The Andes rising above the Atacama Desert of northern Chile on the border with Bolivia are inhospitable and uninhabited. No tree, no shrub, nothing but sparse steppe grass adorns this gigantic landscape crowned by volcanos. Here, at an altitude of about 3,500–4,500 meters, the air is so clear that all colors appear unreal in their intensity. Numerous turquoise and emerald-green lakes are nestled among the giant peaks. Given the solitude and sparseness of this high altitude world, one is surprised to see the lakes carpeted with the rose-pink bodies of flamingos and other birds. Among them lives, alone or in small groups, a very rare bird, the horned coot (*Fulica cornuta*), which is about 50 centimeters long and gets its name from a hornlike protuberance above its beak. When the horned coot (which lives mainly on aquatic plants) gets ready to brood, it tries to build its nest on a small flat island, since the barren shores offer no protection and the hills teem with hungry small predators. The big problem is that natural islands are very rare indeed. And so the horned coot goes about building its own island. It picks an appropriate spot on the water, between 60 and 80 centimeters in size, and proceeds to carry stones there in its beak, swimming to and fro. It drops the stones in such a way as to gradually build up a cone, 3 meters wide at its base, that reaches almost to the water level. Then it tears off some aquatic plants and drags them across the lake to the pad. In the end, the layer of vegetation is 35–60 centimeters thick, about 2 meters at its base and narrowing down to about 50–60 centimeters at the top. This example of planning and building an island residence is, I believe, unique in the world of birds.

Protective packaging

Some marine birds, such as the guillemot, which lives in the northern seas, lay their pear-shaped eggs without the shelter of a nest directly onto the bare windy rock ledge of the rookery. The center of gravity is far from the center of the egg. When the egg starts to roll on the sloping rock, it goes into a curved course and returns all by itself to a position of equilibrium. By its own dynamics, the egg protects itself from falling off. Presumably, all those guillemot eggs that did not have this property were destroyed.

As a package, the egg has a lot of amazing peculiarities. The thin calcium shell is constructed in such a way as to make it almost impossible to crush the delicate egg along its longitudinal axis with one's fingers. Many egg-eating animals have had to learn extraordinary tricks to crack the shells. The African egg-eating snake (*Dasypeltis atra*) swallows the egg whole and saws it open with dagger-sharp cervical vertebrae as it passes through the gullet. The Egyptian vulture (*Neophron percnopterus*) cracks ostrich eggs open by pelting them with stones. Eggshells are fragile only when subject to short thrusts. They do not have the elasticity to withstand strong, sudden impact. There is a good reason for this, for it is precisely this mechanical weakness that helps the fledgling free itself from the shell.

The egg's shock absorption, which has received little investigation, is based on the fact that the embryo is surrounded by the albumen, an elastic gelatinelike substance of high water content. The result is a propitious combination of properties: a liquid that cannot be compressed, only displaced, and an elastic substance. When the embryo is pushed against the shell by some forceful impact, the liquid must flow past it and transform the destructive energy into heat.

The shock absorption of the egg is further improved by an air cushion located at the thick end of the egg—the same end as the center of gravity. In a falling body the center of gravity moves to the lowest possible point, so in an egg the embryo falls on the air cushion. The air pocket in the egg has another mechanical function: It prevents temperature fluctuations from cracking the shell. (We take this into consideration when we pierce the blunt end of an egg before boiling it.)

Nature has provided many packaging methods to protect plants and animals against physical impact. The design often uses a minimum of material while offering a maximum of usable space. Tetrahedral elements which can be stacked in hexagonal containers without any waste of space are frequently found in plants. Beehives too have an optimal geometric design. Their hexagonal shape provides a maximum of inner space for the brood with a minimum of material and no unnecessary interspaces. One could imagine them having been formed by the pressing together of cylindrical cells and the adjustment of the six contact areas to identical thickness. It is notable that the geometric formation is equally ideal at the base, where two layers of honeycombs are put together to form a double comb. Here too material is used economically. The result is a honeycomb floor consisting of three rhombic planes. Each of these planes represents a common wall between the two combs, so that the pattern of cells is shifted one against the other.

The honeycombs of bees can of course not be expected to be mathematically exact, but their precision is quite amazing: The cell walls are built to a thickness of 0.073 millimeter, with deviations of not more than 2 percent. The diameter of the honeycomb is 5.5 millimeters, with 5 percent tolerance. The cells are inclined at a 13° angle against the horizontal plane. Humans would need rather fine measuring instruments for such precision. Bees achieve it with their antennae. Our knowledge of the physical principle by which they measure length and angle is incomplete.

Road construction and traffic control

A mighty tropical storm is an unforgettable experience when it comes down upon you through the green roof of the rain forest. You do not see the approach of the ink-black thunderclouds, but you feel that the sun has disappeared and the wilderness has fallen silent. Then you hear an ominous rumbling from afar. It sounds like a huge animal moving slowly closer. This eerie sound comes from the raindrops hammering on the treetops. Soon you hear it above you and you begin to breathe the moisture of the atomized drops. In a little while the canopy of leaves is soaked through and the drops start battering the soil. Visibility is down to a few meters, and the jungle turns into a grayish-black landscape streaked with white mist. You move on, soaked to the skin, and when you get to a less protected spot the falling water presses you against the ground. Again and again your body flinches under the lightning and the thunder. Ten or twenty minutes later, it's all over. You wade up to your ankles through water and mud and you walk through clouds of intense fragrance wafted down from the flowering trees. Single rays of light are reflected in the myriad of drops. The first bird calls penetrate the jungle.

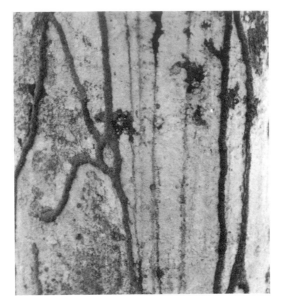

Roofed termite roads on the trunk of a jungle tree in Brazil.

It was from such an experience that I first came to understand why ants and termites of the jungle cover their roads and paths with roofs. Their passageways cling to the bark of giant trees and wind their way up like thin lengths of brown or black hose. They are glued together with endless patience from tiny crumbs of soil and stone. Without such painstaking efforts, the insects would inevitably be swept off by every tropical rainstorm. I have observed with great interest the many embranchments of their complicated road system, and have occasionally opened one of their tunnels to watch their repair techniques.

We certainly are not alone with our traffic problems and our methods of solving them; the teeming populations of ants and termites have them too. We know little as to how far the social insects have progressed in this respect, but a single observation led me to the conclusion that their methods may be highly developed. I noticed a junction on the trunk of a jungle tree where four tunnel roads of a termite species came together. It was not a direct crossing; rather, the roads met in a clearly visible circle—a perfect model of a traffic rotary.

Structure, stability, and power

The importance of size

The largest animal ever to have populated the Earth—more than twice the size of the largest prehistoric saurian—is living to this day in the oceans: the blue whale. This mammal weighs as much as about 40 elephants, that is, approximately 140 tons. Still, its length of 30 meters seems modest compared with the giant sequoias of California, the largest living organisms ever. At their base they reach a diameter of 10 meters, and they grow more than 100 meters tall. Even they are surpassed in height by some eucalyptus trees. At the other end of the scale there are bugs (for instance, the *Ptilinidae*) that, notwithstanding their complicated multicellular structure, are less than a quarter of a millimeter in size. The tiniest multicellular creatures, the rotifers, weigh less than 0.0001 gram.

Despite these fantastic differences, size is not a mere whim of nature. Rather, it is determined by strict laws of physics given by the mechanical structure of a creature, its environment, and its behavior. Active fliers, for instance, cannot exceed a relatively modest size for reasons of aerodynamics, since their flight muscles would otherwise not be strong enough for the liftoff. Muscle strength increases in proportion to the cross section, while weight increases in proportion to the volume— that is, body weight increases faster than the muscular strength gained by the increase. The limits of the ability to fly are almost reached by the condor and the albatross. They are to a large extent dependent on favorable winds in order to fly. The now-extinct pterosaur, which reached up to 8 meters in wingspan, probably could glide only for short distances.

Insects owe their ability to transport a multiple of their body weight and to jump over obstacles that by far exceed their size to the fact that they are so small. Their muscles are extremely strong in proportion to their body weight. Moreover, they can use them at high rates of speed; since the body surface decreases with the square of body measurements, while the volume decreases cubically, they have an excess of surface area for breathing.

The fact that body volume increases faster than body surface also plays an important part in the thermoeconomics of living creatures. The blue whale may have 10 million times the weight of a mouse, but its surface is only 10,000 times as large. Its gigantic size makes it easy for the whale to produce a great deal of body heat; at the same time, it provides efficient protection against excessive heat loss, since the contact surface with the water is relatively small.

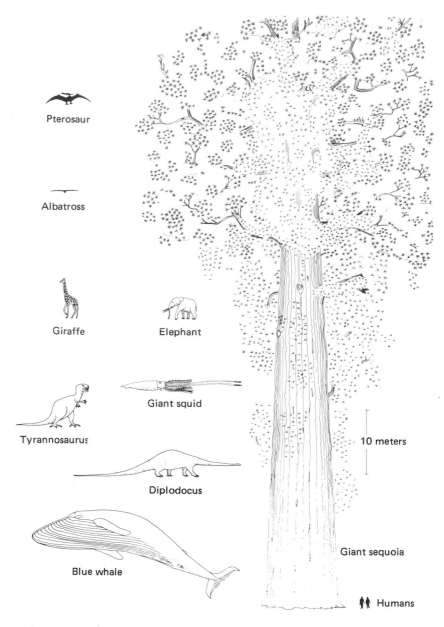

Pterosaur

Albatross

Giraffe

Elephant

Tyrannosaurus

Giant squid

Diplodocus

10 meters

Blue whale

Giant sequoia

 Humans

Relative sizes of various creatures.

Size, therefore, is an important survival weapon. This applies not only in matters of temperature, but also where the intake or expulsion of water and air is of importance. Gravity is one of the factors that determine the sizes of animals and plants. Bones have limited carrying power, and the whales were able to develop only in the water, where buoyancy counteracts gravity. They choke when beached, because their lungs are unable to function under their own weight. The huge prehistoric saurians must, like the whales, have found relief in the water from their crushing weight. Likewise, the longest plant in existence, the seaweed *Macrocystis* of the South Pacific, which grows as long as 200 meters, managed to develop only by floating on the ocean.

Body size is also of aerodynamic importance. An ant that falls off a tall tree can go back up without any trouble, but were the same to happen to a man he'd be lucky to survive without serious injuries. Without air, all bodies would fall at the same speed; but friction increases with the surface exposed to the air current. The friction applied against the fall of an ant is enough to brake its low weight. To small creatures, therefore, the air seems much more viscous than to large ones.

The physical conditions of the environment differ for creatures of different size. It is a good thing that the tellers of fairy tales had no idea of such complications. Dwarfs would have such high-pitched voices that no one could hear them, and they would be sucked in and drowned by the surface tension of the water as soon as they would try to take a drink. Giants would collapse under their own weight or perspire so much that they would have to continuously cool off in the water.

Sturdiness and stability

The giant trees of California reach 100 meters, slender bamboo stalks rise up to 40 meters, and grasses no thicker than a millimeter are sturdy enough to let their long blades sway in the wind. Why do these plants not collapse under their own weight? How are they constructed so they stand up against the force of a storm? Physics can be conquered only by physics. It is certainly worthwhile, then, to look closer into the physical background of strength.

The force required to bend an object of some length to a certain curvature depends very much on its composition. Luckily, we can determine the strength properties with the help of just two independent values which are easy to understand and can be measured separately. One of these values is the elasticity of the material, described by the modulus of elasticity E. Very flexible materials, such as rubber, have a small E value; less flexible ones, for instance bone, have a higher

E value. The resistance of an object against bending increases with its *E* value. The second value does not take into account any property of the material, but its distribution. It is called the moment of inertia of the body, the *I* value. It works as follows. When a thin metal rod is being bent, the material at its inner and outer surfaces needs to be compressed or stretched only a little. However, if the same quantity of metal is distributed over a pipe with thin walls and a larger diameter, the respective distance of compression and stretch is considerably larger. Since more force is required the greater the stretch, it is much more difficult to bend a hollow pipe than a solid rod. That is, the *I* value of a carrying object grows in proportion to the median distance of its material to the axis of flexion. This is the secret of the remarkable strength of a T bar, a cylindrical hollow arm, or a structure of corrugated iron.

The stability of tree trunks or grass blades against bending increases with the product of their *E* and *I* values. It is more difficult to bend them, the less elastic their material and the further away it is from the axis. The pressure they can withstand before cracking or losing their shape also increases proportionally to the value of this product. These laws of the theory of strength are too important for nature to take them lightly. They are carefully adhered to, particularly in designs that aim at providing carrying structures with the highest possible moment of inertia. Many plant stems show economy of material in the interior and a pipelike arrangement on the outside. Some of the most familiar examples are stems of flowers, stalks of grain, and bamboo. Many bones of birds likewise consist of air-filled pipes. The same structural principle is encountered in the shaft of a feather and in innumerable other carrying elements in biology. Thin-walled pipes can effectively be protected against breakage by the insertion of stiff rings. This idea has been realized in bamboo and in the horse willow.

Modern pipes, such as those used for natural gas and oil, often have spiral reinforcements, which are also found in the structure of diatoms, in the trachea of insects, and in conveyor vessels of plants. This method of increasing the mechanical resistance by means of fiber spirals, which are easier compressed and stretched without breaking, is also applied in the spiral growth of tree trunks. For inexplicable reasons, many trees start growing spirally under strong pressure from wind and snow, as is often encountered in the mountains or in subarctic regions. Under such circumstances the wood fibers may deviate up to 30° from the vertical direction of growth. Since this spiral growth offers the plant better protection against mechanical destruction, it may well be regarded as a kind of defensive reaction. The direction

A stalk cut open with a machete, showing the distribution of the woody material that results in greater bending strength (Amazon region, Colombia).

of the twist depends on the species. Horse chestnuts, for instance, always spiral to the right; poplars to the left.

In tropical jungles, with their great variety of species, we encounter a multitude of mechanical ideas for construction. The long thin strands of buttress roots by which the jungle giants brace themselves against the swampy ground are built and grouped so that they can withstand maximal pressure without breaking. The cross-section of the tree sometimes looks like an asymmetrical cogwheel. The projecting wooden cogs, which at times widen at the end into a T, increase the moment of inertia of the trunk and decrease the danger of breaking. We also find thin trunks joining into bundles, supporting each other and forming an upward winding spiral. Obviously, the plants compete for the light at the top by sophisticated technical means.

Nature also applies neat mechanical construction ideas to increase the strength and hardness of flat and bowl-shaped structures. Many insect wings show a stiff corrugated surface. The wing of the dragonfly is set in rigid pleats. Grasshopper wings look like and can be folded like fans. The same fan principle is also used to stiffen the large leaves of many palms. And the shells of many animals are fitted with remarkably sophisticated details to increase their strength. Even the wavy ribbed design of many seashells is nothing but a clever maneuver in the competition for greater strength.

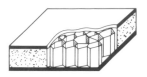

Durvillaea antarctica
(seaweed)

Bird bone

Sandwich structures

Hollow stem   Bamboo   Insect trachea   Cross-section of
                                                              Aspidosperma tree

Structures resistant to bending

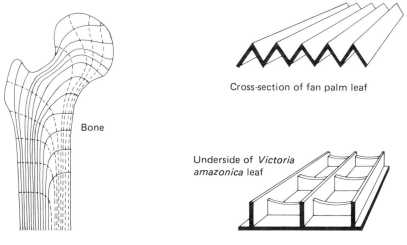

Cross-section of fan palm leaf

Bone

Underside of *Victoria
amazonica* leaf

Half-timber technique   **Surface reinforcements**

In the realm of nature one encounters all the essential elements of structural
engineering and architecture.

Jungle tree trunk with "corrugated" surface and spiral structure (Amazon region).

The pleating in the leaf of the fan palm stiffens the large leaf blades.

Fiber construction in bones and insect shells

Except for a visit to a museum of natural history, there is hardly a chance in civilized surroundings to get to admire the mysterious shapes of bones. Not so in out-of-the-way desert and mountain regions, where only the condor or the vulture pays attention to a dead animal. There one may often find oneself in front of a pile of bones bleached perfectly clean by the sun. It is quite surprising, then, to see how irregular bone structure can be, like abstract sculpture. Only when one is in the mood for a jigsaw puzzle and puts the bones together to form the chain of a spine or of a limb does one begin to realize the sophistication of their mechanical structure.

Bones are designed to offer optimal resistance to tensile and compression stress. Upon closer examination of their delicate tissue, one finds a dense scaffold of the finest osseous beams. They form parallel bundles along the axis of the bone, and they fan out in the condyles so as to hit the bone surface at a right angle. They are aligned precisely in the direction of the compressive forces acting on the bone and can therefore check them efficiently. At right angles to these bundles are others that have the task of intercepting the tensile stress and keeping the bone from breaking apart. These delicate beams of bone serve the same function as framework structures in houses, bridges, or cranes; there too the supporting or linking elements are as far as possible aligned in the direction of the forces brought to bear.

However, nature is much more delicate and elegant. Nature has no reason for making a bone round or square. The outlines of bones, therefore, follow the stress lines or are vertical to them so that they give an indication of the pressures the bone has to withstand. But this ideal distribution of bone material along the stress lines would have been to little avail were the material itself not so well adapted to extraordinary pressure. Just like fiberglass made of synthetics threaded with glass fiber, bone tissue is made up of two constituents which greatly differ in their mechanical properties. About half the bone volume is made up of inorganic crystalline material. It consists of phosphate, calcium, and hydroxyl ions and comes very close to hydroxylapatite in structure. It appears in the bone in the form of tiny crystals, only about 200 atomic diameters in size. They are inserted between thin fiber hairs of the elastic material collagen and seem to be linked with them. Many of these parallel inorganic and organic building blocks form fascicles, which may be interwoven in various ways. The end product is a material that is considerably stiffer than collagen, though low in weight, but by far not as brittle and inelastic as pure hydroxylapatite. Besides, because of the continuous alternation between brittle and elastic material, there is little chance for a fracture to spread unchecked.

When rubber for automobile tires is reinforced by being mixed with tiny coal particles, the resulting mechanical properties are qualitatively similar to those of the material used by nature for bones. We still know very little about the exact molecular basis for the extraordinary physical properties of these compound building materials, which has worked out so well that nature has been using it in many structural elements and not only in bones. Tiny inorganic crystals are interspersed with proteins in shells, teeth, corals, and eggshells. Likewise the hard shells of insects are intricately put together of mechanically very different components, namely chitin and protein. Chitin, which resembles the cellulose of wood, forms long chain molecules held together in very thin fibers. These delicate fibers are linked with each other by proteins, and the shell formed in this way has mechanical properties that differ greatly according to the kind of linkage.

A very similar idea for achieving great strength seems to have been realized in plant cellulose. Very small areas in which the cellulose is crystallized are arranged in regular patterns and surrounded by areas in which crystallization has been prevented. That is, in cellulose too brittle microregions alternate with elastic ones. When we compare modern building materials based on similar physical principles (such as ferro-concrete and fiber-reinforced synthetics) with the meticulous submicroscopic network of biological composites, we must acknowledge that we still have a lot to learn from nature.

Elasticity
The Jivaro Indians who inhabit the catchment basin of the Pastazo River in the Ecuadorian jungle have become known for their custom of killing foes with a blowpipe, cutting off their heads, and shrinking them to the size of a child's head. The bones are extracted through a cut in the back of the head, and the skin is boiled with herbs. Then the lips and eyes are sewn shut with a thread and the head is filled with sand and dried slowly and carefully on hot stones. The sand is gradually removed as the skin keeps shrinking. By means of the heat treatment and the extraction of water, the biologically important elastic material collagen is transformed from an orderly, almost crystalline state into a disordered form. During this process, the collagen fibers shrink to about one third of their former size.

The shrinking technique of the Jivaros shows clearly how important an ingredient of human skin collagen is. Collagen is also a main component of the sinews and ligaments. Neither is it absent in bones and cartilage. What is elasticity's role in nature, and how did it come into being?

When a rubber band is stretched or a willow branch bent, forces come into play which seek to reestablish the original state and the original form. These elastic forces depend on the diameter and profile of the deformed object, and, of course, the material. The effect of the material, as we already know, is described by the modulus of elasticity, which indicates the force needed to produce a certain elastic deformation. Collagen fibers can be deformed 10 times easier than bones or oak wood, but it is almost 1,000 times more difficult to deform collagen than to deform slightly vulcanized rubber or other biologically important elastic materials such as elastin, abductin, or resilin.

Elastin is a component of the arterial walls, and it imparts to them the elasticity that is essential for efficient blood flow. It is also frequently found in larger quantities in the necks of such vertebrates as horses, where its elastic properties reduce the amount of muscular strength needed to hold up the head. Abductin occurs, for instance, in the ligament of the scallop. When the shells close, the ligament is elastically compressed; when the scallop relaxes its muscles, it makes the shells spring open. Scallops are able to produce recoil power for swimming by the rapid periodical opening and closing of their shells. No doubt they owe this ability to abductin. When the shells shut, the abductin stores the energy needed to reopen the shells in its elastic structure. The elastic material resilin finds similarly purposeful application in the thorax of flying insects. The wings of the insect are kept in motion by periodic elastic deformations of the thorax. Resilin prevents the loss of motion energy with every wingbeat at the moment of braking on the point of reversal. Its elastic properties cause the flight mechanism to act like a rubber ball, which retains its kinetic energy when bouncing off the ground though it simultaneously reverses direction.

These examples illustrate one of the most important functions of elastic constituents in biology: to take on and store kinetic energy that can be used to perform work. It is also the elastic materials that supply some animals with seemingly tremendous power. For instance, if the work performed by a jumping flea were to be measured against its muscle volume, the results would be scary. But we know today that the flea applies an elastic mechanism that recalls a slingshot. Before jumping, the insect compresses an elastic resilin structure by slowly moving the muscles of its hind legs. Then the compressed material is suddenly released and the stored energy is freed all at once as kinetic energy. Grasshoppers jump by means of a similar catapult mechanism, but they store the elastic energy inside the set of muscles used for the jump.

Purposeful and elegant uses of elasticity abound not only among animals; elasticity is an equally intrinsic part of plant life. Above all, elasticity makes it possible for plants to avoid a variety of mechanical pressures. Also, it is instrumental in the ejection of seed from dry seed pods by a catapult mechanism. Caoutchouc, the raw material for rubber, was in use in the plant world millions of years before it was introduced into our technology.

There is a considerable difference between the elasticity of metals and the elasticity of organic building materials. When steel is stretched by a few tenths of a percent, it breaks; but soft rubber can easily be stretched to three times its length without tearing. The different elastic behavior is easily explained by the different chemical and structural composition of the two materials. Steel consists of regularly arranged iron atoms. When it is stretched, the atoms move away from each other and the metallic link between them breaks. Rubber has an entirely different structure, a tangle of long elastic molecules which are occasionally linked with each other chemically. When rubber is stretched, its long molecules are simply straightened a bit in the direction of the pull. When the rubber is released, the molecules return to their original nonorderly state. The storing of elastic energy, therefore, is identical with a transformation of the rubber molecules into a somewhat ordered state. The elastic substances resilin, elastin, and abductin have elastic properties very similar to those of rubber, though their molecules are structured differently.

A somewhat different kind of elasticity can be observed in elastic substances such as collagen or in many plant fibers. There the molecules are in part already aligned and ordered, so that stretching mainly causes the distances between atoms and molecules to grow. Therefore, one cannot expect more than a 10–20 percent stretch in fibers of this kind. Nature has varied and adapted the molecular structures of its building materials in such a way as to make them able to cope with the manifold physical demands they have to meet.

Sandwich structures

I spent a stormy winter on the rocky coast of central Chile, where I was working in a small lab. The roar of the huge waves against the steep cliffs was the music I heard continuously, day and night. Whatever man put forth against the power of the sea—wood, concrete, or steel—was destroyed in the course of time. Nothing seemed able to withstand the waves. There was one exception, however: Precisely at the place where at ebb tide the full force of the surf beat against the rock, a lush vegetation of algae was thriving. With each breaker, a storm of movement ran through the jungle of fanlike or snakelike aquatic plants, which reached up to 1–2 meters in length. Their arms whirled and undulated in the suction of the current, but they survived the mechanical stress undamaged for months or years. Why weren't the algae torn off, crushed, and pounded to bits against the sharp rock? Such extreme physical stress could only be conquered by a masterful design. When I first tried to get hold of an alga, I was not able, in spite of the greatest effort, to tear off a single strand no thicker than my finger. Neither could I sever it against an edge of the rock. Only when I cut open the broad arm of a brown alga (*Durvillaea antarctica*) with a razor blade did its secret come to light. It was ideally constructed according to the sandwich method. Its coarse outer skins were braced against each other by a meticulously structured layer of polygonal honeycombs. A light-metal or structural engineer would have jumped for joy at the sight. All the questions I had asked myself seemed suddenly to have found an answer.

Sandwich structures are construction elements in which two thin membranes are linked to each other by a very loosely and lightly built supporting layer. The flexible membranes thus become a mechanical unit which is not only much more stable, but also to a great extent protected against local breaks, cracks, and deformations. The sandwich structure owes all these advantages to the loosely assembled filler, which transmits the mechanical forces and distributes them quite evenly over large areas of the membrane surface. In this way, no stress great enough to destroy the membrane can appear anywhere.

Modern light-metal technology is no longer conceivable without sandwich structures. The narrow wing profiles of fast planes, for instance, are often filled with honeycomblike supports of sheet metal — exactly the same construction method that allows the brown alga to weather the mechanical stress of the surf.

Strong, light cardboard is produced by covering a wavelike inner layer of paper with two flat outer layers. The advantageous properties of plywood sheets are likewise based on the principle of sandwich construction. Nature has made generous use of this opportunity to produce optimal mechanical qualities by the proper distribution of building material. Wherever high mechanical stress must be withstood and at the same time lightness is required, we find sandwich structures: for instance, in the thin walls of grass blades, in the shafts of feather-grass, in the scales of butterflies, and in the delicate calcium structures of marine animals.

This brown alga of Chile (*Durvillaea antarctica*) owes its stiffness and hardiness to its sandwich structure.

The structure that braces the two membranes of the sandwich is only in rare cases a simple honeycomb like that of the brown alga. In most cases the connecting walls have numerous additional supports, so that the filler between the membranes looks like foam. These supports may consist of the same material as the membranes. In principle, the advantages of the sandwich effect can be produced simply by the economic use of the building material in the interior of a mechanically stressed body. This sophisticated saving method is encountered in the bowl-shaped bones of the skull (important in terms of mechanical protection), in the shells of lizards, and in the horned beaks of birds. A special case of sandwich elements are tubular structures, whose walls are additionally braced against each other. The weight-saving tubular structure is thereby combined with the mechanical advantages of the sandwich structure. Impressive examples are the tubular bones of birds, which are often efficiently reinforced by filler structures. Many plant stalks too are built according to this principle. Like the elder, they have a hard tubular outer wall and softer foamlike pulp inside. A similar structure is found in the spines of many plants and animals, which are subject to a lot of mechanical stress.

Wherever the destructive effect of a strong force is to be efficiently checked, it must be splintered into many small forces, which disperse and cancel each other out. The compressive, tractive, and bending forces must not be allowed to grow so strong in one spot that the material is unable to withstand the impact and breaks. Nature has succeeded to the point of perfection in doing just that for structures under mechanical stress.

Muscle power
Amid the noise of engines and the clouds of exhaust gases, it is easy to forget that there are also noiseless and clean ways to transform the chemical energy of fuels into power and motion. One need not search long for an example; nature succeeded in creating such an ideal mechanism in the form of the muscles.

The "gas" of the muscle machine is a small, energy-rich molecule called adenosine triphosphate (ATP). Manufactured in specialized chemical factories of the body cells, ATP supplies by disintegrating the mechanical energy by which a muscle is contracted while performing work. The secret of power production can only be understood if we are able to shed light on the structure of the muscle elements which are movable with respect to each other. In a certain group of muscles with a distinctive striped pattern the structure was indeed decoded. Two kinds of protein fibers interlace like fingers, and the pattern is regularly repeated along the axis of the muscle (thereby producing the stripes). The thin fibers are called actin, the thicker ones myosin. Depending on whether or not these protein fibers overlap to a greater or a lesser extent, the muscle is contracted or relaxed. When the actin and myosin fibers are stretched to the extent that they no longer overlap, muscle strength is no longer produced. On the other hand, the more overlapping occurs, the more power can be produced. It is, therefore, very probable that the transformation of energy and the power action originate in the many little "feet" which protrude from the myosin and bridge the gap to the actin, an average distance of about 70 millionths of a millimeter. According to the most popular theory in this field—that of Huxley and Davies—it can indeed be assumed that these tiny appendages change their structure under the influence of ATP and perform thereby a specific movement. This would simply mean that the myosin fibers act like chemical centipedes crawling along the actin structures. In other theories (those of Elliott, Rome, and Shear), it is proposed that electrostatic and molecular forces produced by ATP cause actin and myosin to telescope and that the bridges between them exist in the first place for the purpose of maintaining the gap between the protein fibers. In other words, in spite of the ample and valid experimental research into muscles, the decisive step in the transformation of chemical energy into mechanical energy has not yet been explained.

There is no doubt that nature has managed to create here a highly sophisticated biophysical mechanism. Could it compete with modern power engines? The conversion of energy in muscles under optimal conditions is 30–35 percent efficient. This efficiency corresponds to that of the combustion engine. The muscle fibers of the crayfish can develop a tractive force of 7.45 kilograms per square centimeter. It follows that a giant muscle with a cross-section of one square meter would be able to lift almost 75 tons.

The fascinating mechanical applications that muscles offer to animated creatures are possible only in connection with solid organic structure elements, such as bones or insect skeletons. Nature developed a multitude of sophisticated joints with which parts of the body can be moved in relation to each other by muscle power. These biological mechanisms have many construction features that are well known in mechanical engineering. One of the most remarkable muscle-moved mechanical elements is the kinetic chain that appears in the skulls of many birds and reptiles and in the mouth structures of many fishes. A kinetic chain is a row of links connected with each other by movable

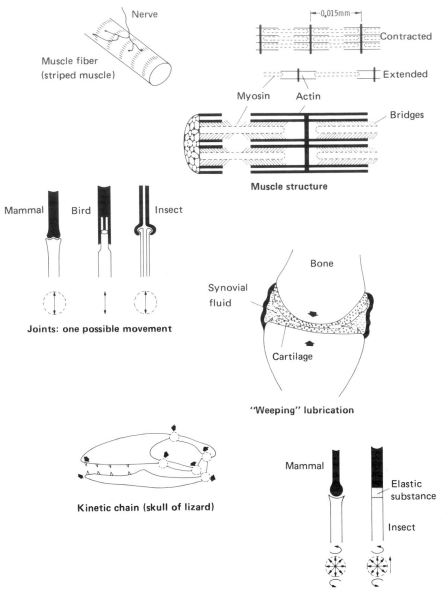

Muscle structure

Joints: one possible movement

"Weeping" lubrication

Kinetic chain (skull of lizard)

Joints: many possible movements

The muscles still guard the secret of their power. Power transmission by joints and lever mechanisms has many parallels to mechanical engineering, but the principle of "weeping lubrication" is exclusive to nature.

joints in such a way that, when one link is held in a fixed position and another one is shifted, all the links move in a fixed course. In this way a simple contraction of a muscle results in a complicated, elegant, and precisely timed mechanism of motion. This principle has found wide use in technology (for examples one need only think of the lever system, the typewriter, the locomotive, or drawing instruments).

Lubrication

Can the lubrication of mechanical bearings be basically improved? Engineers working in this field will probably shake their heads "no." Today technology is able to reduce the sliding resistance of metal bearings to less than 1 percent of the weight the bearing carries. Surfaces can be polished until they are smooth as glass, and the innumerable lubricants offered for sale hardly differ in quality. Such perfection notwithstanding, it cannot be excluded that lubrication technology has entered a dead-end street.

Nature has chosen different ways to cope with the problem. The natural bearings most dependent on good lubrication are the joints, where bones are movably connected. They not only need to withstand great stress, they also must be safeguarded against attrition and their inner temperature must never approach 50°C lest the precious proteins suffer damage. A superficial observation of a knee or hip joint shows nothing that differs from similar technical arrangements. The area of the joint in which the spherical condyle and the bowl-shaped socket touch is enclosed by a membrane which forms the "oil sump" containing the synovial fluid that lubricates the contact surfaces of the two bones. What is special about these joints is the quality of the gliding surfaces.

Whereas technology tries to polish the hard metal of bearings to as fine a finish as possible, nature covers the touching surfaces with a spongelike substance which is comparatively stiff yet quite elastic: cartilaginous tissue, which differs from hard bone tissue essentially in that it lacks deposits of calcium crystals. One could compare this tissue to fiberglass from which the fibers have been removed. The fine pores of the cartilaginous sliding layer are soaked through with lubricating synovial fluid. When the joint is subjected to pressure, the layer compresses and the fluid is pushed out of the thin ducts. The gliding principle is the same as the one used for air-cushion vehicles, with the difference that in bone bearings the cushion (of fluid) is produced on the spot.

To study this mechanism of "weeping" lubrication, Lewis and McCutcheon passed a hard rubber sponge soaked in a soap solution across a glass surface. The friction resistance it produced was only three thousandths of the weight it carried. It hovered, so to speak, above the fluid pressed out of the cavities in the gliding surface. Sliding bearings of this sort can obviously compete with conventional mechanical bearings.

Synovial fluid might also compete in the market with modern lubricants. It contains slightly less protein than blood serum, but on the other hand it carries an organic acid with very long molecules which are probably linked to proteins. The more the gliding speeds vary in the lubrication layer, the lower the viscosity of the fluid. Very slow gliding results in viscosity about 5,000 times that of water, very fast gliding in only about 10 times water's viscosity. In this way pressures are checked more efficiently and distributed from areas of great stress on the bearing to areas of less stress. Modern lubricants aim at the same properties but lack the elasticity of synovial fluid. Another remarkable feature of this fluid—one that counteracts the attrition of the gliding surfaces—is that it acts like a thin layer of soft rubber and cannot be entirely displaced by pressure on the stress points of a bearing.

Tools

Archaeologists and anthropologists classify ancient cultures and races according to the state of development of their tools. The quality and the material of tools determine how well raw materials can be worked. In the construction of tools, attention to the most minute mechanical detail pays off. A tool is generally tailored precisely for its function.

Nature has produced many tool devices that are amazingly similar to our modern tools. For example, the idea of the pliers appears in all its variations in the biting apparatuses of many mammals, reptiles, and insects. Lobster claws transmit power by using leverage. The strong beak of the woodpecker works like a chisel. Wood wasps and ichneumon flies have sophisticated drilling tools which, though they do not rotate, penetrate the wood just as well by a forward motion. Leeches have saw blades of a very hard material, with which they work their way through the skin of their prey by steady sawing motions. The tongues of many mollusks are built exactly like rasps and act as such in cutting food into little pieces. The horny tongue of the woodpecker has sharp barbs and functions like a fish hook or a harpoon, pulling insects out of cracks in the wood. Among the teeth of predatory

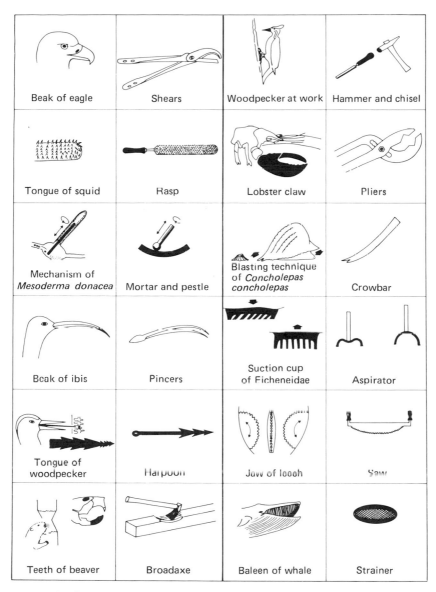

Beak of eagle	Shears	Woodpecker at work	Hammer and chisel
Tongue of squid	Rasp	Lobster claw	Pliers
Mechanism of *Mesoderma donacea*	Mortar and pestle	Blasting technique of *Concholepas concholepas*	Crowbar
Beak of ibis	Pincers	Suction cup of Ficheneidae	Aspirator
Tongue of woodpecker	Harpoon	Jaw of loaah	Saw
Teeth of beaver	Broadaxe	Baleen of whale	Strainer

Basic tools of nature and technology.

fish we find a wide assortment of blades and knives. A carnivorous marine snail of Chile (*Concholepas concholepas*) has two strong protruding teeth on the lower rim of its solid shell, which it wedges under the shell of its prey clinging to a cliff. By pulling its foot in, making use of the crowbar principle, it can then easily detach the prey from its base. Pincers of all kinds and cleverly constructed tongs can be recognized among the exotic beaks of birds. Nature uses suction devices such as we use for clearing clogged sewers, and suction plates, as holdfasts. The baleens of whales operate like strainers, to filter krill out of the water.

What degree of hardness is nature able to produce in its tools? Layers of magnetite, an iron oxide, are found on the teeth of the tongue (radula) of sea snails which feed on rock-clinging algae. This compound is close to number 6 on Mohs's hardness scale from 1 (talc) to 10 (diamond), which means that it is just one position below quartz. In other words, nature works with materials that do not lag behind our strongest steel.

There seems to be no end to the range of tools nature comes up with. The often very numerous versions of the same tool point out the importance of the finest structural detail in the struggle for survival. The problems encountered by a bird when taking hold of a slippery fish are considerably different from those involved in picking up a kernel or sifting through mud. Tools in the animal kingdom are fascinating products of a selection that often involves rigorous physiotechnical criteria.

Physical technology among tool-using animals

I was observing a small flock of gulls hunting for marine animals along the rocky shore of a fjord near Puerto Natales on the southwestern tip of Chile. Nothing in their behavior seemed out of the ordinary. I paid hardly any attention to a single bird climbing gradually higher into the air. Suddenly I was startled by a sharp plopping sound. A small object burst into several smaller pieces on the rocks below. The gulls swooped down upon the small pieces and found tasty morsels inside. Examining pensively the remains of the shell which the high-flying gull had shattered on the rocky ground, I thought of Newton and man's long hard quest for an understanding of free fall.

Gulls probably learn from experience that the forces released at the moment of impact by a falling shell are destructive. As flying creatures, they surely have a fine sense of gravity. What is amazing is that they instinctively stop using their beaks under certain circumstances and proceed to mobilize gravity for a specific purpose. Their skill, however, is greatly surpassed by the African vulture (*Neuphron percnopterus*). This vulture, generally not very choosy with respect to its prey, employs special skills when looting a nest. Like the gull, it makes use of gravity when the blows of its beak fail to open an egg; but since ostrich eggs apparently are too heavy to be lifted, the vulture uses stones. Carrying a stone of 100–300 grams, the bird approaches a nest, grips the stone with its beak, raises itself to a vertical position in front of the egg, and hurls the missile down onto the shell.

The sea otter of California (*Enhydra lutris*) often swims on its back and uses its chest as a tray to deposit and clean its catch of marine animals, or as a playground for its young. It has devised a special trick for cracking hard shells: It places a flat stone on its chest and bangs the shell hard against the stone. This technique would be futile when tried directly against the soft elastic chest.

Some animals seem to be instinctively familiar with the laws of ballistics of moving objects. The lasso spider (*Mastophora*) skillfully aims a sticky pellet attached to a long thread at passing insects. The archer fish (*Toxotes jaculatrix*) shoots a spray of water the distance of up to ten body lengths at insects sitting on plants above the water.

Neither has the usefulness of the lever remained a secret from animals. The North American nuthatch (Sitta pusilla) uses thorns or thin pieces of wood to poke insects out of galleries, seeing that its beak is too short for the task.

There are numerous reports about the use of tools among primates. Chimpanzees, for instance, use sticks both as an extension of the arm and for hitting

3

Locomotion

The struggle against friction

Whether an animal moves over the earth's surface, flies through the air on wings, or plows through the water, it must fight the forces of friction which are working to brake its movements. The amount of frictional drag depends on how the air or water flows around a body in motion. The dynamics of flowing liquids are subject to complicated physical laws — laws which life has learned to master in minute detail in the race for higher speeds and greater savings in energy. Torpedo-shaped fish profiles, spindle-shaped bird bodies, and the narrow builds of fast land animals are nature's concessions to the physics of flowing media.

When water or air circulates around a body at a certain speed, a thin layer is formed on the surface of the body which acts as a brake on water or air molecules. The molecules on the surface itself are stopped altogether. In the direction outward their speed increases gradually until, at a distance of a fraction of a millimeter, they reach full fluid velocity. The gradual and steady decreases of fluid velocity toward the surface of a body, where water and air particles adhere, is the result of interactive friction exerted by the flowing liquid or gas layers. Therefore, the more viscous the fluid, the thicker this boundary layer of divergent fluid velocities. The type of flow that develops, and the generation of eddies and turbulences and therefore of drag, are essentially determined by the conditions in this boundary layer. The flow is described as laminar if the air or water particles flow steadily and parallel to each other; it is turbulent if the molecules drift every which way and many small eddies are formed at irregular intervals. In a turbulent boundary layer a great deal of energy is transformed into heat through collision and friction, at the expense of kinetic energy.

Turbulent boundary layers thus exert a considerable braking effect on motion.

Whether the flow around a surface is laminar or turbulent depends primarily on the condition of the surface, the viscosity of the medium, and the velocity of the flow. One cannot simply compare flow conditions for objects of similar shape but different size. The flow around two objects of the same shape but of different length can be compared only if they have the same Reynolds number. The Reynolds number is obtained by multiplying the length of the object, the fluid density of the friction-producing medium, and its fluid velocity, and then dividing the value thus obtained by the viscosity. A fish about 10 centimeters long swimming at a speed of 10 centimeters per second will have a Reynolds number of about 10,000; the same result is obtained for a fish 2 cm long moving through the water at 50 cm/sec. A 10-cm bird flying at 140 cm/sec has the same Reynolds number as a small dragonfly moving at 1,500 cm/sec. The astonishing thing is that the configuration of flow around these animals would be directly comparable if the streamlines of their bodies were similar.

The lower the Reynolds number, the more stable the laminar boundary layer. But as soon as it reaches about 3,000,000 on a flat surface, turbulence begins and the energy loss through friction is greatly increased. At what point turbulence sets in for any given object depends on the pressure pattern along its surface, as well as on the presence of edges and projections from which instability can spread. Frictional drag also increases at very low Reynolds numbers (under 100), since at this level the viscous forces of the flowing medium take effect.

Besides the Reynolds number, shape is of decisive importance for the total drag of an object in a flow field. To understand this we need only imagine a fish swimming in the water. Pushing head first, it must part and displace the water, and to this end it struggles against pressure drag. As the water flows by the body of the fish, the flow process in the boundary layer produces friction drag. Then, flowing together again at the tip of the tail, it forms eddies that use up energy, again at the expense of the kinetic energy of the fish. In order for the fish to lose as little as possible of its energy to friction, its frontal surface must remain small. The pressure along the surface up to its largest diameter must decrease gradually, so as to prevent the flow from piling up and becoming turbulent. The formation of eddies because of rough shapes and edges must be avoided as much as possible. If all these complicated requirements are coordinated, the result is the ideal torpedolike streamlined shape. Nature has come very close to this ideal structure in the body shapes of numerous animals, especially the fishes.

A streamlined body of optimal shape has a length of 4.5 times its diameter. In this case the surface is smallest relative to the volume. This optimal numerical proportion has not remained a secret from nature: for dolphins (*Tursiops gilli*) the ratio is close to 5.

At low and very high Reynolds numbers, laminar flows are generally no longer advantageous. If the Reynolds number is high, the following happens: Because of the strong pressure gradient at the point of confluence toward the tail end of the streamlined object, the laminar boundary layers tend to separate from the surface and produce extensive paths of turbulence. This phenomenon of separation occurs much later when the boundary layer is turbulent, since there is sufficient exchange of energy between surface and flow. A compromise can be reached if the largest diameter of the streamlined body is pushed far back so that the area in which turbulence forms and loss of energy occurs is much more limited. As a matter of fact, the maximal diameter of the dolphin is slightly back of center. Obviously, the above-mentioned effect was taken into consideration. Experiments have confirmed that this particular shape lowers friction drag in turbulent boundary layers to 65 percent.

Locomotion in water

Fluid-mechanical tricks of rapid swimmers
We keep discovering refinements of fluid mechanics in the animal kingdom. Recently Rosen and Cornford investigated the effect of the mucous secretions of fish on turbulent friction drag in seawater and found that 5 percent of barracuda slime reduced the turbulent friction of seawater by 66 percent. The slime of halibut was found to have a similar effect.

These results indicate that slime in the flow boundary layer of rapid swimmers can effectively subdue turbulence and thus prevent energy loss. No doubt many observers of nature have long since suspected something of the sort. A new field of inquiry has thereby been opened in flow mechanics for the purpose of understanding and controlling this effect. Experiments that would explain the physical effect of fish slime in more detail have not yet been done. It is, however, not difficult to understand the process in principle, bearing in mind that the Reynolds number of a flow decreases with the increasing viscosity of a fluid. Assuming that mucous secretions (such as castor oil) are about a thousand times more viscous than water, it is simple to estimate that a few percent of the stuff in the boundary layer can easily lower the Reynolds number to a tenth or less. In this way the danger of turbulence can be considerably diminished.

Slimes consist of polysaccharides, chain molecules composed of hundreds and even thousands of carbohydrate elements. The longer molecules reach a thousandth of a millimeter in length, and would be visible under the microscope if they were not so thin. They increase the viscosity of the water by interacting with water molecules through weak intermolecular forces (adhesive forces) and thereby greatly restrict their freedom of movement. This explains why the appearance of microscopically small eddies can be suppressed on the mucous bodies of fish.

Instead of mucus, which must be secreted continuously, nature may just as well have attached thin threads—hair, that is—to surfaces in a flow field. As a matter of fact, the short wiry hair or feather coats of such good swimmers as seals and penguins seem to be a great advantage for the maintenance of laminar flows. This is borne out by technical experiments with fine wire on surfaces in a flow field (M. D. Kramer, 1938), as well as by indirect clues. The splendid fur of the seal, for instance, provides no protection against getting wet; it gets soaked through in the water. And, when exposed to the air, wet fur presents a considerable risk of a chill. We must assume that wet fur represents an advantage for swimming. And, like the seal's coat of hair, the feathers of birds may have a favorable effect on the boundary layer.

Of very special interest from the viewpoint of fluid mechanics are the dolphins. About 2 meters in length and with top speeds of more than 8 meters per second, *Tursiops gilli* easily reaches Reynolds numbers of 10 million, at which the normal streamlined body already shows turbulence. But because the largest diameter is located back of the body center, the turbulence-prone surface of the dolphin is reduced to the tail area. This clever feature, though, does not by itself explain the dolphin's extraordinary swimming performance. Measured by their muscle weight, dolphins simply swim too fast. Examination of their smooth hairless skin reveals a spongy layer in the outer area bordering on the surface of the skin. This layer is 1.5 millimeters thick and consists of interconnected cavities filled with a fluid. When pressure fluctuations (such as occur in turbulent boundary layers) are transmitted to these cavities, they subside and are distributed by the back-and-forth movement of the fluid, much as in a waterbed. Eddies such as those that appear behind body protuberances or those generated by turbulence could be subdued in this way (Kramer 1960, 1965).

Mackerel with prey (sardine) in its mouth. Note torpedolike body and sickle-shaped tail fin.

By examining separately the body shapes, the boundary layers, and the movements of dolphins and other rapidly swimming creatures, we shall probably never quite grasp their fluid-mechanical achievements. Their secret lies in the meticulous coordination of many hydro-mechanical and dynamic factors.

Harmony of shape and motion among fishes
Anybody who has visited a large aquarium has a vivid idea of the fantastic variety of shapes to be found among fish. As manifold and complex as their shapes are, at first sight, the mechanics of their swimming motions. What common physical denominator could be expected among the dancelike movements of a large-finned toyfish, the slithering motion of an eel, the powerful short fin strokes of a tuna, and the flightlike movements of a ray?

In the course of biological development over millions of years, structure and movement adapted to the very different environments and ways of life of the various species. Not all fishes must be rapid and tireless swimmers in order to survive. Some travel freely and unhurriedly through the oceans; others dart anxiously through the jagged maze of a coral reef, looking for cover. By carefully examining the many shapes and movements and including what we know about the

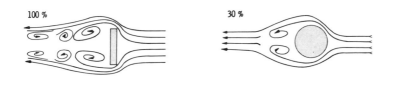

Flow resistance

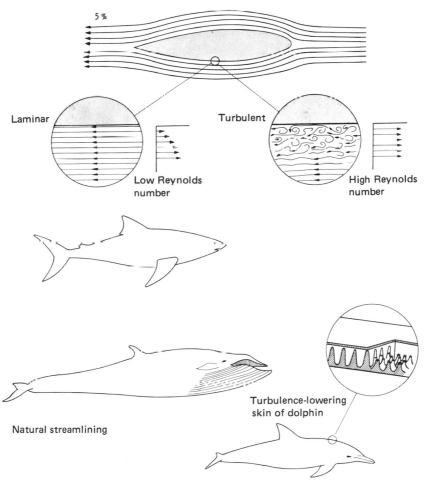

Laminar

Turbulent

Low Reynolds
number

High Reynolds
number

Natural streamlining

Turbulence-lowering
skin of dolphin

Through millions of years of natural selection, optimal flow shapes and
sophisticated flow aids have developed.

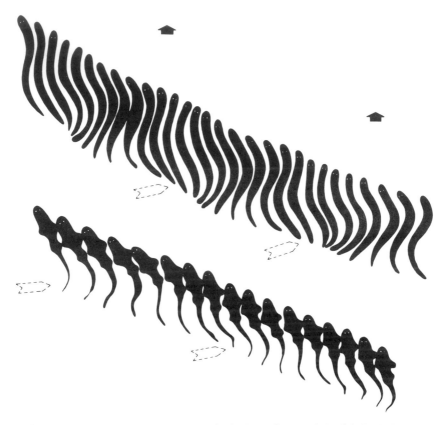

The periodic waves that run through the bodies of spotted dogfish (*Halaelurus chilensis*) and eel-shaped lampreys (*Heptatretus politrema*).

evolution and development of the species, we find a connecting thread on which to hang some first physical speculations.

The most primitive and ancient method of locomotion among water dwellers was probably the lateral wriggle, whereby a wave travels from head to tail and increases in amplitude. Many primitive invertebrate swimmers use this kind of locomotion. In many aquatic animals (lampreys, eels), wriggling is aided by vertical stabilizer fins that extend the sides of the body and thus facilitate power transmission to the water. Cat sharks, some true sharks, lungfish, and sturgeon also swim with this lateral slithering motion. How efficient and energy-saving this method of locomotion is can best be seen by taking a look at the eels, which cover thousands of kilometers on their wanderings through the oceans. Slow-motion pictures of their movements disclose the principles of physics on which they are based. The essential prerequisite for getting ahead by wriggling through the water is for the body wave to travel rearward faster than the fish travels forward. The wriggling animal thereby exerts pressure on the water along the wave loops moving backward. While the laterally directed components cancel each other out over the entire fish, the forward- and backward-directed components add up to the propelling force.

A simple but significant modification of shape and motion leads from this pure slithering locomotion to another widely used swimming mechanism. The front of the fish remains immobile and the wriggle is limited to the last third of the body, with the amplitude of the wave starting from almost zero and rising sharply toward the tail. This modified wavelike swimming motion restricted to the tail area is used by a great number of species. Typical representatives of this technique are trout and salmon, but it is also used by porgies, red mullets, barracuda, horse mackerel, sea bass, and perch. Why has evolution so favored this motion? On the basis of research in fluid mechanics, we know today that this kind of motion is more effective because it imparts to the displaced water greater acceleration than does a pure slithering motion. Energy-consuming eddies thus have less time to develop and to brake the motion of the fish.

To be sure, nature had to overcome a serious physical drawback in adopting this improved method of propulsion. Whereas in a slithering motion that runs along the entire body the lateral forces are well canceled out, they cannot be absorbed when only the tail area oscillates. So that the entire body of the fish would not be tossed from side to side by recoil with each flip of the tail, certain preventive measures had to be introduced. The lateral cross-section of the body was enlarged and extended by vertical fins. The lateral forces produced by the swing of the tail, which cause the adverse recoil, were reduced to a minimum. Nature intervened precisely where the forces are strongest because of the high acceleration (power = mass × acceleration)—namely, where the wave motion most increases. With the specific purpose of reducing recoil, the cross-section of the tail roots became considerably smaller, without noticeably affecting the propelling effect of the oscillating tail. The constriction between the main body and the tail is especially marked in tuna, swordfish, and mackerel.

In the tuna, adaptation has taken an additional step. On either side of the constriction, streamlined cartilaginous protuberances further reduce the flow drag on lateral swings. The fastest swimmers of the oceans—tuna (*Thunnus*), wahoo (*Acanthocybium*), striped marlin (*Tetrapturus*), louvars (*Luvarus*), sailfish (*Istiophorus*), and swordfish (*Xiphias*)—have developed a striking crescent-shaped tail fin. (The same shape has also evolved in sharks and whales, so we must assume it to be of considerable advantage for rapid swimming.) The velocities reached by fishes in this group are indeed extreme. Yellowfin tuna (*Thunnus albacares*) have been measured at 20.8 meters per second. Wahoo (*Acanthocybium solandri*) have been clocked at 21.4 m/sec (77 km/h, or 42 knots), three times as fast as the first atomic submarine, the *Nautilus*. The frequency of the tail movement can exceed ten full strokes per second.

The secret of the sickle-shaped fish tail has not yet been completely unraveled. One suspects it lies in the ideal structure of the eddy street in its wake. Though the eddies produced by the tail swings consume energy, they supply the necessary recoil for the forward motion. What matters is that a high momentum (mass × velocity) be imparted to the eddies and that a minimum of energy be wasted. Smoke rings which can be blown through the air for quite a distance are one example of such energy-efficient eddies. In the wakes of rapidly swimming fish, nature has probably come close to an ideal eddy street.

There may be additional fluid-mechanical advantages to the sickle tail. A close look at the tail fin of a tuna shows that the front of the sickle is broad and rounded, while the rear comes to a very fine point. Each stroke of the tail forces the fin against the water's resistance, whereby the thin rear part deforms to give the tail the shape of a wing set at an angle against the direction of locomotion. As a consequence, additional driving forces appear which periodically thrust the fish forward. In view of the deflectability of the oscillating "wing," this driving principle could well be compared with the whirr of the hummingbird.

A number of good swimmers among the shark and sturgeon families are heavier than water. If they did not mobilize vertical forces, they would slowly but inevitably sink to the bottom. Since they are continuously in motion, nature was able to solve their weight problem in a very elegant way: The sickle-shaped tail is asymmetrical. Because the upper half is larger than the lower, its resistance produces upward-directed torque. In addition, the pectoral fins are shaped like small wings and function as elevators, producing and controlling vertical forces.

Pectoral fins first served as steering devices. In fish families that have no need for high velocities, they have been transformed into rowing organs. The swimming motion of the wrasse is restricted to simultaneous strokes of the pectoral fins. The ever-active inventiveness of nature did not fail to gradually improve the efficiency of the pectoral-fin stroke. These fins grew larger and stronger until water resistance

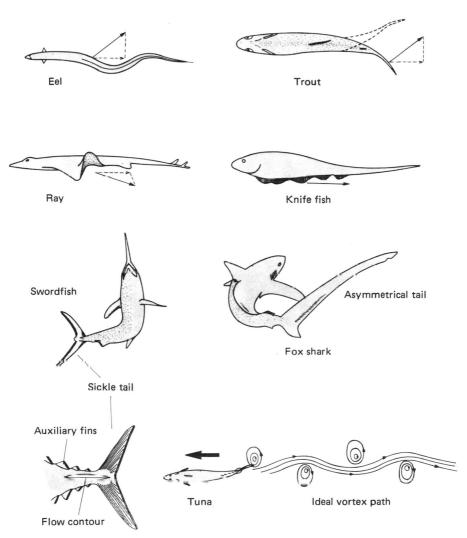

Hydromechanical means of propulsion.

set a limit to the stroke. The rays of the *Raiidae* family, which consist almost entirely of pectoral fins, swim by sending a periodic wave over the fins. Thus, they have come full circle back to the original undulating driving mechanism—with the difference, though, that the wave motion is now vertical instead of lateral. Actually, at this point nothing would have been in the way of further development, by reduction of the wave, to the rowing technique used by birds in flight. And the fact is that rays of the family *Myliobatidae* have developed a structure that flaps just like a wing, whereby the downstroke is essentially responsible for propulsion. Other body fins too have widened enough to be used for propulsion by wavelike motion. The North American bowfin uses for this purpose its long dorsal fin, the electrical eel its long anal fin.

Thus the development of fishes was decisively influenced by the wriggle, the original method of locomotion in water—or, rather, by the simple laws of physics to which this method owes its effectiveness. One could almost say that fish shapes represent a variety of solutions to a physical problem nature was faced with, given a variety of initial and marginal conditions.

The jet principle

Rocket and jet engines are based on the principle of recoil. The generation of recoil is the result of a basic law of mechanics, the law of the conservation of momentum (momentum = mass × velocity). This law says that in a self-contained system the resulting momentum is conserved in magnitude and direction. If, for instance, someone dives from a small boat into the water, the boat receives momentum in the opposite direction. This momentum corresponds to the mass of the diver multiplied by his kickoff velocity; the heavier the diver, the higher the velocity at which the boat is thrust back. Rocket engineers try to keep the exhaust velocity of hot gases as high as possible, which, in accordance with the law of conservation of momentum, increases the speed of the rocket.

Nature did not bypass the jet principle. When the larva of the dragonfly species *Anisopterus* is frightened, it pulls its legs up to its body and strongly contracts the water-filled rectal chamber in its rear end, expelling a strong current of water which propels the larva about 10–15 centimeters. Moving at a rhythm of about three jet impulses per second, it can reach a top speed of 30–50 centimeters per second. Large larvae (*Aeschna, Anax*) effect the contraction of the rectum in one tenth of a second. The escape velocity of the water reaches about 23 cm/sec.

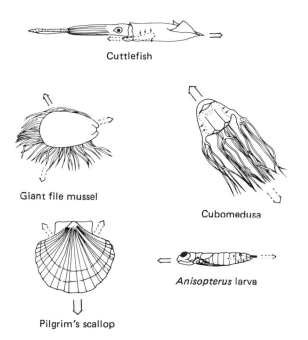

Cuttlefish

Giant file mussel

Cubomedusa

Anisopterus larva

Pilgrim's scallop

Jet propulsion has been discovered in nature repeatedly and independently.

The most important group of living creatures using the jet principle are the *Cephalopoda*, including octopi, nautili, and squid. They all are able to expel water under high pressure from the cavity (pallium) through the so-called funnel. That is how they swim along, moving in jolts, their tentacles extended. Some octopus species are capable of using recoil to shoot out of the water and glide some 15 meters through the air.

Some remarkable and unexpected recoil swimmers are found among the scallops (*Pectinidae*) and file shells (*Limidae*). When such a bivalve is approached by a predatory starfish, it instantaneously slams its shells shut. The water that escapes at the margin of the mantle generates recoil and the bivalve moves away, hinge forward, in the opposite direction. By rapidly repeating this movement, bivalves are able to swim by jolts, changing direction all the time. If, however, they are not in too much of a hurry to get away, they mostly swim with the shell aperture forward. In this case the pilgrim's scallop (*Pecten jacobaeus*) expels recoil-generating water on either side of the hinge. File shells can, while swimming, expel water not only next to the hinge, but also in other directions (particularly downward). Thus they can determine the direction of their recoil motion and perform what amounts to a dance in the water.

A surprisingly well-developed jet technique is found among the otherwise rather primitive medusas and jellyfish. The ectoderm of the hydromedusa envelops the bell cavity in the form of an annular diaphragm whose opening can be controlled by muscle fibers. While swimming, it can thus regulate the speed of the water expelled by the contractions of the umbrella. The bell cavity of the fire jellyfish (*Cubomedusa*) is likewise enclosed on the bottom by a similarly operating and controllable circular membrane that concentrates the jet stream. The jet is guided by the bases of the tentacles, which have developed into strong lamellae that can be extended from the bell cavity to serve as steering organs during swimming. By contracting the umbrella two or more times per second, this nimble creature is able to reach continuous speeds of 3 kilometers per hour and a top speed of 9 km/h.

It is notable that nature has applied the jet principle only in water. This makes perfect sense, because the recoil generated is proportional to the mass of the expelled medium. The acceleration required to generate effective recoil with gas would be too much for biological conditions.

Hydraulic mechanisms

Because fluids cannot be compressed, they transmit power reliably without causing any elastic deformation of containing walls. Since fluids can be carried conveniently in pipes of whatever shape, they are excellent means of power transmission for many difficult mechanical problems. Hydraulic mechanisms have proved extraordinarily useful in motor vehicles and other machines. Whether nature has learned to use fluids for power transmission is a most interesting question.

One need only walk through a garden to catch sight of an admirable hydraulic mechanism: the earthworm. Wherever it contracts its longitudinal muscles, it gets short and thick. In each small segment of the body the sphincters and the longitudinal muscles operate alternately, and in each segment the contraction occurs somewhat later than in the one in front. As a consequence, crosswise- and lengthwise-directed waves of contraction run alternately through the body to the rear and push the worm forward in the direction of locomotion. The greatest pressure which earthworms can generate and transmit in the fluid chambers of their bodies corresponds to the pressure exerted by a one-meter-high water column. Many mollusks (animals without inner or outer skeletons) obtain and control their shape through hydrostatic pressure. Sea anemones use the flexibility of their hydrostatic skeleton to greatly expand or reduce their size according to need. Many bivalves activate hydraulic mechanisms to bore themselves into the sandy bottom by alternately pushing their shells together and apart through fluid movement and hydrostatic pressure.

A remarkable and effective hydraulic mechanism is found in the legs of spiders, which have muscles to flex the joints but none to extend them. Spiders stretch their legs by pumping fluid into them. When a spider gets ready to jump, it generates, for a fraction of a second, excess pressure of up to 60 percent of an atmosphere. The legs extend in order to accommodate more fluid.

In arteries, nature has developed a brilliant mechanism for power transmission. Were the walls of arteries not elastic, blood would be transported intermittently; it would stop moving between contractions of the heart. But since arteries are elastic and capable of expanding with each heartbeat by up to 10 percent, they store elastic energy with each beat. The elastic recoil of the vascular walls then propels the blood also in the intervals between heartbeats, and thus provides continuous blood circulation in spite of the rhythmic pumping.

In the plant kingdom, hydrostatic mechanisms have been developed to catapult seed. The squirting cucumber (*Ecballium elaterium*) of the Mediterranean regions gradually builds up an excess pressure of several atmospheres (about the same pressure as in a truck tire) inside the fruit, until the seeds shoot out through a pluglike stopper that bursts open. The initial velocity can reach 50–60 kilometers per hour, and the range is close to 13 meters. This is a considerable achievement, as the seeds are ejected at an angle steeper than the optimal ballistic angle of 45° so as not to get caught in neighboring plants. The North American dwarf mistletoe (*Arceuthobium*) flings its seed out of the fruit by a similar catapult mechanism at 90 km/h and propels it almost 15 meters away.

How well nature has learned to deal with the physical properties of fluids is demonstrated by the African and Indonesian cobras that can direct two thin jets of venom through openings in their teeth at the eyes of an enemy from a distance of several meters, and by the snapping shrimp, which can numb its prey with a sharp jet of water.

Flotation mechanisms

Proteins, an intrinsic part of living substances, make up about 16 percent of the body weights of fishes. Since protein substance is heavier than seawater, ocean creatures are faced with two choices: In order not to sink, they must either swim continuously or possess a flotation mechanism. Only by getting down to a body weight that corresponds to the weight of the displaced seawater will they stop sinking.

Wherever there is a question of radical economy measures, it is essential not to overlook inconspicuous factors which, when added up, may become very important indeed. Obviously, nature has in the first place taken under close scrutiny the weight of salts contained in body fluids. Seawater contains 35 grams of salt per kilogram of fluid. Body fluids too contain salt, which, when dissolved, disintegrates into ions which are indispensable for the transmission of nerve impulses and many other life processes. Two methods of saving are possible: reducing the concentration of salt to a minimum, and replacing ions of high atomic weight with ions of lower weight. Nature has experimented in both directions.

In many fishes, the salt concentration is considerably lower than in seawater. As a consequence, their eggs rise in the water, which improves their odds for survival. Water has the tendency to dilute salt and flow into the sea. It must therefore be contained within watertight walls if low salinity is to be retained. Precisely for this reason, nature prefers in many cases to maintain the salt concentration of the sea inside the living organism and merely reduce the ionic weight. Large quantities of ammonium chloride are found in some octopi (*Cranchia scabra*). The ions of ammonium weigh only half as much as those of potassium, and the ions of chlorine are much lighter than those of phosphate or sulfate. By agglomerating ammonium chloride, an end product of protein metabolism, these octopi become lighter than seawater at the rate of 16 grams for each liter of body fluid.

The device of turning waste products into means of floating is not only ingeniously simple, but also very economical. Among marine animals, however, fats, which can be imbedded in many body tissues, are more widely used as buoyancy generators. Fats have a lower density than water and are extremely light in sharks and whales which dive to very great depths. For these animals air bladders and pulmonary cavities do not present a viable mechanical solution, because of the high water pressure. Fat, on the other hand, is not only light in weight but also difficult to compress. Even at great depth, it maintains almost unchanged buoyancy.

Air is about a thousand times lighter than water and appears therefore to be an ideal filler for floating devices. About half of all osseous fishes use the principle of the gas-filled air bladder for altitude control, as do some species of insects, mollusks, protozoa, and algae. The mastery of its dynamics, however, is anything but simple. If the walls of the air bladder are elastic, the gas pressure inside the bladder

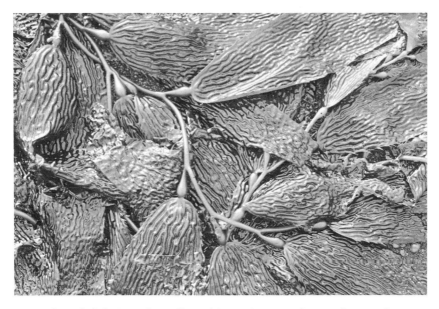

Seaweed needs light in order to live. This species stays close to the water's surface with the help of gas-filled air bladders. Its light antennae remain fanned out because of their corrugated structure. (Straits of Magellan, Chile.)

increases together with the outer water pressure. At a depth of 2,000 meters the pressure exceeds 200 atmospheres. As a consequence, the gas volume is reduced to the 200th part. The remaining gas space contains air that has become 200 times denser. It follows that the buoyancy of the air bladder has practically disappeared. Furthermore, gas that is so strongly compressed dissolves quickly in seawater, in which low oxygen and nitrogen concentrations are held in balance with the air above the surface. Only an active mechanism capable of carrying gas molecules into the air bladder against a steep concentration gradient can maintain its function in great depths.

The mechanism of this gas pump is still somewhat of an enigma. Air bladders do not always contain air. Close to the water's surface, nitrogen often predominates; in great depths it is oxygen. There even exists a species of fish (*Physalia*) in which carbon monoxide is found. Nature successfully uses air bladders in depths far exceeding 2,000 meters. Depending on the water pressure, gas is either pumped into the air bladder or eliminated by it. This complicated control system and the continuously changing size of the air bladder have their disadvantages. That is why some cephalopods (such as *Sepia officinalis*) have developed another kind of floating device: the cuttlebone (*ossa sepia*), which contains a great many tiny cavities in a regular pattern. These are mechanically stable because they are supported against the outside, and contain mainly nitrogen. No matter how deep the octopus dives, its pressure always remains at 80 percent of one atmosphere, so that the creature always remains in balance with the nitrogen dissolved in seawater. To control buoyancy, the octopus simply pumps water out of its skeleton and allows gas to fill the emptied cavity. In principle, the cavities of the cuttle bone function just like water tanks in a submarine.

Land locomotion

When the first living creatures rose out of the primal sea to conquer the dry land, they were faced with new mechanical problems. Their means of locomotion, adapted to the laws of flow, proved to be of little use when faced with the unpredictable and irregular conditions on the earth's surface. In the course of evolution, these means of locomotion were thus either replaced by others or, if sufficiently flexible, modified and elaborated for multiple-purpose use.

The latter happened to the original wriggle. When snakes move over loose pebbles or cross a waterway, they still move exactly like eels and other undulating marine animals. But this simple horizontal serpentine motion does not meet the snake's many other needs. Additional mechanisms had to be developed. When a snake crosses a furrow that is too narrow for horizontal wriggling motions, it makes its body undulate vertically like water waves. When it crawls through a hole, rhythmic contraction waves run lengthwise along its body. On rough terrain, a snake has the option of folding its body into accordion-like pleats and moving by stretching itself and then folding up again. All these clever techniques, however, fail miserably in very loose sand, which may be burning hot to boot. What else can a snake resort to, when the soil no longer affords enough of a hold to push off? What device would an engineer propose?

Desert-dwelling snakes have hit upon the sophisticated technique of winding sideways. They combine horizontal and vertical undulations so skillfully as to produce two rolling contact areas with the sand so that they can push the body forward between the two as far as the next point of contact. All that remains of their passage is a parallel double track in the sand, the imprint of the rolling touchdown spots. Convincing proof of the effectiveness of this locomotion can be found in the fact that it was developed independently by desert snakes of three continents: the horned viper of the Sahara (*Cerastes cerastes*), the

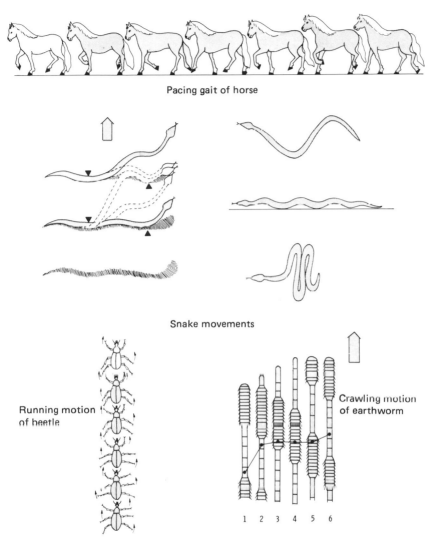

Pacing gait of horse

Snake movements

Running motion
of beetle

Crawling motion
of earthworm

1 2 3 4 5 6

Mechanisms of locomotion on land.

American rattlesnake (*Crotalus cerastes*), and the saw-scaled viper (*Echis carinatus*) of central Asia.

For getting ahead in narrow earth tunnels, the hydrostatic mechanisms found in many worms have proved successful. Here propulsion is effected by contraction waves which run backward along the body segments and thereby stretch the body forward. Hydrostatic forces also come into play in the progress of snails. Contraction waves move rhythmically from the rear to the front of the creeping disk and thereby push the clumsy body forward over sticks and stones.

On land, however, animals would not have progressed very far with such technically remarkable but slow and awkward methods of locomotion. The ultimate breakthrough came with the development of legs. The great physical advantage of legs lies in the application of the elegant laws of leverage to locomotion. Now short muscle movements were able to produce large movements of the body. Lifting the body off the ground reduced friction resistance greatly. Creatures were suddenly able to move easily over the bumpy surface of the earth.

A daddy longlegs can scamper to safety across a maze of obstacles with amazing speed. While worms and snails progress no more than a few meters per hour, the cheetah and the ostrich reach velocities of 100 kilometers per hour. Agile jumpers like the kangaroo (*Macropus*) can, with a runup of only one meter, jump over obstacles 2.7 meters high. (Man's record jumps reach more than 2 meters, although the technique requires lifting the center of gravity only about 1.2 meters.) There are tiny mice (*Zapus*) that can jump 3.7 meters with practically no runup. In a rough world without roads, there was no alternative to the lever mechanics of legs. The inhospitable ground of some faraway planets will, in all probability, also be marked some day by the footsteps of walking robot craft.

When a friend in Lima mentioned once that there were horses in Peru that ran like camels, my face must have reflected my incredulity. He stubbornly stuck to his statement, and soon thereafter I found myself on a stud farm south of the capital. The horses bred there did indeed amble like camels, moving both legs of the same side simultaneously. The normal gait of the horse—the trot, whereby one front leg and the opposite hind leg move simultaneously—was completely unfamiliar to these animals.

I was told by the breeders that these pacers were descendants of horses from Andalusia introduced to the coastal desert regions of Peru by the Spaniards in the sixteenth century. They were used for the long rides between the widely scattered oases, rides that were very uncomfortable because of the loose sand. It became evident that a few individual animals managed to move over the sand much better than the others. Those were the ones given preference for stud, and they became the ancestors of the present Peruvian pacers.

The habit of moving at the rocking pace of the camel is inborn, but the style can be improved by intensive training. Pacers offer a remarkably gentle ride, almost free of the rhythmic jolting typical of normal riding. This must have been the reason why great pains were taken in the Middle Ages to train so-called palfreys as pacers for the use of ladies and clergymen. Because of its smoothness, the pace is much better suited to riding over sand than the trot. It causes an animal to sink its feet much less deeply into the ground and to move faster. Not only desert dwellers like the camel have learned to use the advantages of pacing; it was also adopted by a species of Icelandic ponies for traveling in soft snow. This shows how fine details in the application of mechanical laws can influence the survival chances of animals.

Why did nature fail to invent the wheel?
The wheel has shaped the modern world so decisively that it has become a symbol of progress and technology. No wonder it is considered one of the most brilliant and trailbreaking inventions of man. It is all the more admirable in that it seems not to have been modeled after anything in nature. That nature did not come up with the principle of the wheel has often been presented as proof of the superiority of the flexible human mind over the rigid ways of evolution. Be that as it may, it certainly would be interesting to find a plausible explanation for the absence of the wheel in living organisms. Such an inquiry could be helpful in pinpointing the differences between our mechanization and the technical development in the course of evolution. In order to assess the significance of the wheel as an element of technology, it seems opportune to ask up to what level a culture can get along without the wheel. Is a somewhat developed technology possible without the wheel?

The wheel was discovered in the Old World some 5,000 years ago by the Sumerians. For thousands of years it was used by highly developed nations like the Egyptians and the Greeks and at the same time also by primitive barbarians. This leads to the conclusion that the invention of the wheel is not necessarily the starting point of rapid technological development. In this connection it is of special interest to note that sophisticated civilizations arose in the New World in which the principle of the wheel (in the classic sense) was never put to use.

The Maya, who more than a thousand years ago built huge cities in the Yucatan and achieved impressive technical and scientific feats, were well able to forgo the help of the wheel.

The temples and observatories of the ancient Mexican nations and the fascinating rock cities of the Incas offer an impressive image of what a civilization can achieve without the wheel or the axle—with one reservation. The builders of those cultures knew very well the energy-saving principle of the roller. To transport huge boulders for their stone structures, they made wide use of stone or wood rollers placed under the load. Thus they could easily have developed the wheel—and perhaps they did, since an occasional wheel has been found on a child's toy. The probable reason why the wheel did not take hold in the New World was the absence of strong draft animals.

A locomotive technique based on the wheel would not have been practical in nature, for it is a characteristic of the wheel that it can turn in the same direction for any length of time. Supplying a living wheel with liquid nutrients, with nerves, and with sense organs would have presented unsurmountable problems. And nature simply does not have provisions for spare parts and maintenance.

Still, would it not have made sense to apply the principle of rolling to the transport of loads? When an object is pulled along a base, energy-consuming friction must be overcome. These forces are proportional to the weight pressing on the base, and depend on the nature of the friction-producing surface. If the object is round and can be rolled, friction is considerably lower, since any existing unevenness can easily be overcome. The microscopic hills and valleys on the contact surfaces engage like cog wheels. Rolling friction too increases with the weight of the object, but it decreases in reverse proportion to the radius of the wheel. That is why the largest possible wheels are chosen for driving on sandy ground. It makes little mechanical sense to roll very small loads over uneven terrain. That is why insects drag food and building materials in a rather awkward manner. One exception is the burying beetle. The load it has to transport exceeds its body size and is, in absolute terms, so considerable that the switch to rolling is worthwhile. This beetle shapes the dung into which it will lay its egg into a ball, then rolls it with its raised hind legs backward into the planned burial site. This complicated instinctive behavior shows that nature has not overlooked the advantages of rolling.

Rotary elements in nature

It stands to reason that in the absence of the proper repair and main-tenance services it would not have been practical for animated creatures to grow limbs that would roll on axles. There would be no way of feeding nutrients into the wheel through a rotating axle, or of eliminating the toxic waste products any living organism produces. It follows that the wheel cannot be given a function of life, and that nature could have produced rotating elements only if they were lifeless like the wheels of our cars.

Indeed this is what happened. Bivalves have in a separate gastric sector a quite firm but movable rod, the crystalline style, which is set in motion like a rotary beater by a jungle of hairlike filaments in the stomach wall. In the bivalve *Mesodesma donacea* of the Chilean coast, this style is about 2 centimeters long and 2 millimeters thick and serves mainly to break up the skeleton structures of swallowed plankton. It functions like a mortar and pestle which can be both rotated and pounded. Here nature has produced a rotating element able to transmit power—an astonishing feat.

A motor-driven rotating locomotive mechanism of extraordinary elegance has only recently been discovered in bacteria. The experiments that seem to have definitively confirmed the first known natural driving mechanism based on true rotation were published while this book was in preparation.

The best-documented bacterium, *Escherichia coli*, has a bean-shaped body about 0.002 millimeter long and 0.001 millimeter thick. In water it travels up to 20 body lengths in a second. Compared with this, a man would have to swim at about 120 kilometers per hour. No fish reaches such extreme speeds. But the scientists who were researching locomotive mechanisms were not much troubled by such speed. They discovered under the electron microscope six threadlike appendages mostly twisted together into one flagellum. It was obvious—and could be filmed under strong microscopes—that the bacteria propelled themselves by undulating motions of their flagella. Such flagellar movements are known from much larger monocellular and multicellular creatures and are based on the same principle of power transmission as the wriggly swimming motions of the eel. The eerie thing about the flagellar motions of the bacteria was that the flagella do not contain anything capable of facilitating a regulated and synchronized undulating movement. They are only about 0.00001 millimeter thick, and are composed of linked protein molecules (the flagellin). When the chains are broken up by ultrasound, the molecules reunite by themselves into flagella, similar to what happens in a thread-shaped crystal.

With a special microscope under which swimming bacteria could automatically be held in the field of vision (H. C. Berg), surprising evidence was found that the flagella of bacteria rotate around their own axle like a ship's screw. The flagella as such were not recognizable under the microscope, but it was possible to make their movements visible. Tiny pellets were treated with a biological substance which made them adhere to the flagella so that the visible movement of the pellets reflected the movement of the flagella. It was shown that the pellets rotated around the invisible axle of the flagellum, while the body of the bacterium rotated in the opposite direction. This is possible only if the flagellum rotates around its base. Bacteria evidently have a rigid helical flagellum with which they wind their way through the water. Because of the particular form of the spiral, angular kinetic energy is transformed into directional locomotive energy.

Bacteria seem to have no difficulties in reversing the direction of their tail spiral. Further electron-microscope research has even discovered the rotor disk to which the torque is transmitted, at that end of the spiral which is anchored in the body of the bacterium. The molecular mechanism which generates the rotation of the tail spiral has so far not been explored further. Several mechanisms are possible from the energy viewpoint. One of the most fantastic concepts in biology has come true: Nature has indeed produced a rotary engine, complete with coupling, rotating axle, bearings, and rotating power transmission.

Astonishing feats of primitive peoples

Many examples cited in this book show how the most complicated physical laws have again and again been applied in nature without ever having been understood. By experimenting patiently, discarding poor solutions, and investing a lot of time, nature often has found solutions for physical problems which are much more elegant than what we humans, with all our intellect and our hard-earned knowledge of physics, have been able to come up with. But impressive examples of physiotechnical achievements without science have also been provided by ancient cultures and primitive peoples.

These feats include the moving of many gigantic loads of rock—unique among them a beautifully cut stone base weighing 132 tons that lies close to the ruins of Tihuanacu on the Bolivian *altiplano*, not far from Lake Titicaca. Comparison of rock material has established that this giant rock, called *Puma punku*, was hauled about 12 kilometers. Today a tender for the transportation of a 132-ton rock would hardly find a bidder. What vehicle could carry a weight equal to that of a herd of 40 elephants or of 130 passenger cars? How was it possible for such loads to be moved, by men who knew nothing of the wheel, over rough terrain in the thin air at an altitude of almost 4,000 meters?

When a load is not carried but dragged along the ground, the required forces amount to only a fraction of the weight. But even if friction resistance along the tracks had been greatly reduced by the use of moist mud, human muscle power still had to "shoulder" roughly 20 percent of the load. Counting 50 kilograms per man, more than 500 men would have been needed to haul the *Puma punku*—and more than double this figure on slightly ascending stretches. It is difficult to imagine how so many arms could so much as get hold of the load. Long hemp ropes would not have been of much help where even steel might have failed. These thoughts lead to the conclusion that the principle of rolling was applied. As we know from mechanical engineering, resistance can be reduced to, ideally, 2 percent by switching from gliding to rolling bearings. By placing rough stone balls or stone cylinders under the load and rolling them along a strong enough base, the required expenditure of energy would certainly have been reduced by another 10–20 percent. Thus the transportation problem would have been brought within a reasonable range. On my visits to the Tihuanacu region and also to Easter Island, I tried to find clues to the actual application of rolling transport techniques. The people of the Tihuanacu culture certainly had the necessary skill. On Easter Island, where hundreds of giant stone figures were moved, I found on three sites thick stone slabs into which grooves had been cut. Round stones the size of a fist would easily have fitted in. Did the inhabitants of Easter Island discover the principle of stone ball bearings? From the point of view of physics, it is difficult to imagine any other possibility.

Another fascinating example of physical technology without scientific understanding is the boomerang. The stone-age nomads of the Australian steppe and desert regions discovered with the boomerang a weapon, a hunting tool, and a toy with flight properties so odd they can give even physicists a headache. The boomerang is a contoured, angled wooden throwing stick which is held like an axe with the concave side aimed at the target. It is hurled almost horizontally, with as strong a rotation as possible in a plane perpendicular to the ground. The boomerang flies 10–15 meters straight ahead and then veers to the right or the left, depending on its construction. At the same time it rises and traces a wide arc back to the thrower, floating back to him from behind with a smooth horizontal rotation. No two throws of the boomerang are identical, and boomerangs differ from each other in numerous ways. Often their movements are more complicated than those described here. In general, the trajectory of boomerangs

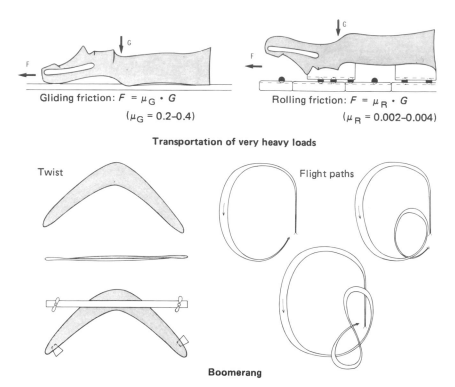

Gliding friction: $F = \mu_G \cdot G$

$(\mu_G = 0.2 - 0.4)$

Rolling friction: $F = \mu_R \cdot G$

$(\mu_R = 0.002 - 0.004)$

Transportation of very heavy loads

Twist

Flight paths

Boomerang

The transportation of the Easter Island statues and the invention of the boomerang are two human feats that were performed without corresponding scientific understanding.

is circular, with a diameter of about 20–50 meters and altitudes of 15–30 meters. They have never been intensively used as weapons, but are well suited to drop a single bird out of a rising flock. The astounding thing about the boomerang is not only that it returns to the thrower when it has missed, but also that its initial vertical rotation changes in the course of the flight into a horizontal end rotation. Anyone who has tried making a rotating disk or top veer from its plane of rotation has been reminded of the law of physics which says that a rotating body seeks to conserve its torque. How then is the flight of the boomerang possible?

The secret of the boomerang does not lie in the line of its curvature, nor in the contour of its profile; both can vary greatly. Rather, it lies in the fact that the planes of the two arms are very slightly twisted toward each other, as if one had tried to twist the boomerang into a helical shape. The Australian natives usually impart this gentle twist to the boomerang over a fire; it can also be produced by filing or by

The entire underside of this fish (*Gastromyzon*) and its fins form a suction plate by which it adheres to a rock.

clamping (as shown in the figure). The theory of the boomerang is complicated. Not only must the wing properties be taken into consideration, but also the forces that, in the course of the flight, act periodically on the unequal rotating arms.

Coping with slippery surfaces

In the rocky surf of the Chilean seashore one often sees men poking around with long iron hooks. One of their favorite catches, and one that requires a great deal of skill, is the *Gastromyzon*, a strange but tasty little fish that has adapted in an extraordinary manner to the dangerous turbulence of its environment. Whenever the surf starts rolling seaward, the fish instantly adheres to a rock, and it remains in the same position when the heavy waves again beat against the cliff. In the brief interval it hunts for prey. While it is glued to the rock, the fishermen try to quickly spear the fish with their barbed hooks and pry it off the rock. The fish's ability to adhere very efficiently to the smooth, slippery rock surface is due to a large suction plate on the underside of its body. Intricate suction devices of this kind are remarkably widespread in the animal kingdom. Starfish, also frequently found in the surf, are even able to move about with the help of their numerous suction cups.

The structural principle of suction organs consists in body inden-
tations with an elastic margin that attaches evenly and tightly to a
smooth surface. The adhesion is produced by a vacuum in the space
enclosed between the adhesive organ and the surface. This requires
energy, which can be released indirectly through elastic deformation
of the suction cup when it is pressed down, as for instance in the case
of toy arrows. Alternatively, the vacuum can be produced by direct
muscle contraction, whereby the interior of the suction device is ex-
panded. If the suction organ is well built, the resulting distribution of
forces adds to the insulating capacity of the margins. In nature, those
suction organs predominate in which the vacuum is produced by
directed muscle movements.

In the well-developed suction organ of the octopus, adhesion works
as follows: The inner surface of the suction cup is turned inside out
against the contact surface. The insulating margin is pressed to the
surface with great strength by contraction of a sphincter. Finally, muscular action creates the inner space of the suction link. The greatest
connecting pressure that can be produced in this way corresponds to
the atmospheric pressure outside; that is, in general, 1 kilogram per
square centimeter. Only the sea anemone *Cribrina* seems to approach
this figure. In small biological suction cups efficiency hardly exceeds
70 percent, in large ones not even 50 percent. Suction cups occur
either singly or in groups, or, if very much extended and subdivided,
in plates.

Nature's imagination is rich when it comes to the application of
suction cups. In the rapids of the Andean rivers there live several fish
species (*Loricariinae, Pygidium, Plecostomus*) that cling to stones by means
of suction cups in the lip area. They let go of their rocks only briefly
in order to make swimming excursions. Frogs that live in running
water have also developed similar suction zones. Other fishes (*Echeneidae,
Remilegia*) cling to sharks or whales with suction plates at the top of
the head. Several tropical bats (*Thyroptera*) suspend themselves by suction
cups on smooth branches or rocks. Monkeys such as *Cercopithecus* and
Inuus use suction devices on their hands to climb very smooth trees.
Cercopithecus need not even put its arms around a tree; it just presses
its palms flat against the surface. Small African mammals of the genus
Hyrax are able to scale vertical or even overhanging rocks with the
help of suction devices. The primate *Tarsius spectrum*, which resembles
the lemur and which scampers about at night in the bamboo of Asian
islands, has suction cups at the tips of its fingers. Suction cups are a
well-known means of aggression and hunting for octopi and polyps.
The larvae of some gnats (*Blepharocera, Liponeura*), which live in torrential

rapids, are able to creep across the rocks for several meters by synchronized use of their six suction cups. The leech too gets ahead by alternately lifting and displacing its two suction cups.

Because the adhesive effect of a suction cup is proportional to its surface and is limited by atmospheric pressure, small, light creatures with their relatively large surfaces can derive greater advantage from the use of suction cups than can large heavy animals. However, this rule does not seem to apply to the giant deep-sea octopi (*Architeutidae*), which reach 20 meters in length and have suction cups up to 15 centimeters in diameter. Sperm whales have been found with deep scars from the suction organs of this octopus on their coarse skins. The apparent contradiction is easily explained. At a depth of 1,000 meters, to which sperm whales dive to hunt for octopi, the water pressure is about 100 atmospheres. Under such conditions suction cups may be 100 times more efficient than at the surface of the sea This is probably why deep-sea octopi grew to such sizes.

Bristled surfaces

Sealskin and pile are indispensable to cross-country skiers. They are attached to the running surfaces of the skis in such a way that the short bristles are directed rearward. This permits easy forward gliding, but in any slip backward against the direction of the bristle the motion is quickly braked when the stiff tips get imbedded in the snow or ice by pressure.

For the seals of the icy seas and the penguins of the antarctic shore this gliding and braking mechanism is vital. When they leave the water to climb an ice floe, they get a grip with their stiff fur or feathers. They can scale 60° inclines in this manner. When they rush back into the water, they simply slide down on their stomachs. Numerous other animals that live in the snow have bristles on their running or creeping surfaces, for the same mechanical reasons.

Soles equipped with bristles or hairy pads are also suitable for locomotion over loose sand. Many desert and steppe dwellers walk on such soft and comfortable soles; notable examples are the tarsiers,*Tenebrionidae* and *Assilidae*, the Ehigmodontia mouse, the sand cat, and the fennec fox.

Flight

The physics of flight
The secret of flight lies in a simple rule of physics perceived in 1783 by the Swiss mathematician Daniel Bernoulli, who studied the pressure

exerted by a liquid flowing through a pipe on the containing walls and found that the pressure decreased with the increasing velocity of the fluid.

A wing's asymmetrical profile forces air to flow by the lower and upper sides at different speeds. The air molecules flowing by the top side of the wing are thus compelled to cover a longer distance and move faster than those flowing by the bottom, if the continuity of the flow is to be maintained. According to Bernoulli's law, the acceleration of air above the wing generates a vacuum whereby upward suction — the main reason for lift — is created. The air flow beneath the wing generates excess pressure by the braking of air particles, which adds to the suction effect above the wing and thus adds to the lift.

Numerous mechanical factors affect lift. The profile of the wing, (particularly its curvature) and the size of the wing surface are very important. The larger the surface exposed to air flow, the stronger the lift. Lift also increases with the speed of flight, since the pressure difference at the wing increases with the speed of the air flowing by. A further factor is the wing's angle of incidence against the current; the pressure difference at the wing increases with its inclination, since the steeper angle causes the speed difference between the flow above and that below to increase.

Because lift is generated only when air flows around the wing, there is no way of avoiding friction drag conditioned by the shape and surface of the wing. In addition, there is drag induced by the air currents and vortices which flight produces. Therefore, in order to maintain flight, the energy lost to friction must be continuously replaced by the propulsion system.

The key to flight — the asymmetric wing profile — was discovered by nature and applied independently in five different orders of animals: flying reptiles, birds, bats, flying fishes, and insects.

The flight of birds

The birds' conquest of the air is due to the extraordinary aerodynamic and mechanical properties of feathers. Feathers have developed in the course of evolution from restructured reptile scales to the most remarkable lightweight constructions in the animal world. They combine great hardiness with astounding elasticity and flexibility, besides shaping the bird body into its harmonious streamlined form.

Feathers differ greatly in size and shape, but they all have a solid, stiff, pointed shaft fringed on both sides by the fine web of the vane. The vane consists of many hair feathers aligned side by side and held together by smaller ramuli which carry barbs. It is because of these

A Galapagos buzzard in flight.

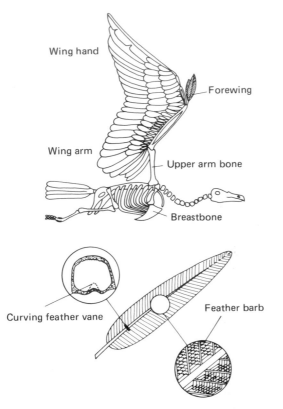

Birds owe their astounding flying capacities to the highly developed mechanical and elastic properties of their lightly constructed feathers and wings.

A heron in flight (Guatemala).

small anchors that the ruffled feathers of a bird can be rearranged in their proper order by a few strokes of the beak. The scape, made of keratin, is a small masterpiece of construction technique: It is hollow at the thick end and filled with a foamlike material everywhere else,. according to the principle of sandwich construction. The concentration of building material in the outer walls of the shaft was necessary to obtain great resistance against bending stress, but nature did not stop here. The walls feature uneven thickness and an indentation on one side which can be folded under stress. As a result, the feather gives when the air current presses against the broad side of the vane, but does not give when stress is exerted in a parallel direction.

Wings developed from the forelimbs of reptiles. Numerous bones fused or regressed, but three digits were preserved, and these play a special role in flight. Two of them were strengthened until they now serve as holding and steering device for the largest flight feathers. The third digit has retained its mobility to a certain extent and, together with a few adherent feathers, forms the forewing (alula) of the bird. Wing arm, wing hand, and feathers work together harmoniously to dynamically adapt the shape of the wing to the most complex aeronautical conditions.

Birds attain considerable speeds in horizontal flight. The peregrine falcon (*Falco peregrinus*) reaches 200 kilometers per hour, the common swift 150. The swallow, the frigate bird, and the golden plover easily break 100 km/h.

The Reynolds numbers of bird wings lie at times far above 100,000 — not very far below the Reynolds numbers of airplane wings. Obviously, bird wings, just like plane wings, had to be built so as to obtain as regular and laminar an air flow as possible. Only thus could strong lift and low friction drag be ensured. In wing adjustability, nature is far ahead of today's technology, for which even the simple "swing wing" was quite a serious technical challenge. Nature, however, was faced with the necessity of building the drive right into the wing. With each flap the wing functions simultaneously as wing and propeller. The shape of the wing changes continuously during each beat. Its dynamics not only reflect the muscle movement of the bird, they are also determined by the flexibility and pliability of the feathers.

The wing arm and the wing hand of the bird serve two different functions that complement each other in the wing beat. The aero-dynamic adjustment of the wing arm near the body remains relatively constant during flight. This is where a great part of the lift is being generated, while the lower wing arm and the wing hand produce the crucial propulsive forces by an oarlike movement. The individual phases of this movement are illustrated in the accompanying diagram (based on slow-motion photographs) showing the flight of the gannet from the side, from below, and from the rear. From below, it becomes evident that the wing hand is taken far forward during the downstroke of the wing so that the wingtips reach the level of the bird's head. Looking at the same phase of flight exactly from the rear as the bird is flying away reveals that, whereas during all other flight phases the wing contours appear only as a thin line, the downstroke shows their lower broadside. What this means is that, during the forward-directed downstroke of the wing, the bird tilts the profile of the wing hand downward at a slant. The result of such a maneuver must be a forward-slanted lift which can be divided into the thrust of the wing beat and additional lift. Toward the end of the downstroke the wings are quickly brought rearward and slightly pulled in. This backstroke gives the bird additional thrust. While the wing is again raised in preparation for the next downstroke, the wing hand, now passively carried along, remains slightly bent so as to prevent any unnecessary drag. After the strenuous start and during long-distance flight, the bird generally refrains from gathering additional thrust during the backstroke and limits aerodynamic propulsion to the downstroke of the wing.

The physical occurrences around a bird wing are, because of its added dynamics, much more complicated than those around the rigid

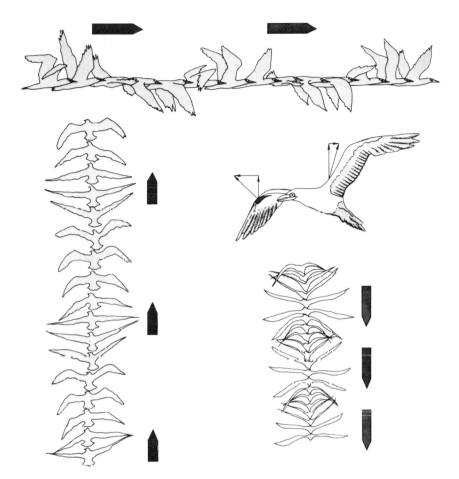

The flight of a gannet, viewed from the side, from below, and from the rear, reveals the secret of the oarlike movement: During the downstroke (arrows), the wing hand is tilted forward and downward, and the lift supplies thrust.

A gull in flight. The light shining through the feathers reveals their careful arrangement.

wing of an airplane. However, many of the mechanisms on which flight is based are the same and can easily be observed, especially when the bird is gliding. For instance, if the wing is too much inclined against the direction of flight to generate sufficient lift at low speeds, strong turbulence can arise above the wing and the even flow can be interrupted. As a consequence, a flying body would lose its lift and pitch down. Aeronautics developed several devices to delay the interruption of air flow, one of which was to attach a forewing above the wing front to guide the air flow toward the top side of the wing. The forewings or alulae of birds serve exactly the same function during slow flight or landing maneuvers. Small protuberances at the leading edge and the surface of the wing can also delay the interruption of the flow by creating a thin layer of turbulence that increases stability. For the same purpose, some birds raise the finest feathers on the surface of the wing.

A widely fanned-out bird's tail, if placed in the right position, can likewise improve lift during slow flight. In certain birds it functions like an airplane's fowler flap (a slotted flap that can be extended from the trailing edge of a wing). In their ability to regulate to a certain extent the curvature of their wing profile, birds also possess a remarkable mechanism for controlling lift. Just before alighting, a bird deliberately allows the air flow at the wing to be interrupted. The consequence is clear to the eye: the wing feathers are ruffled by the turbulence.

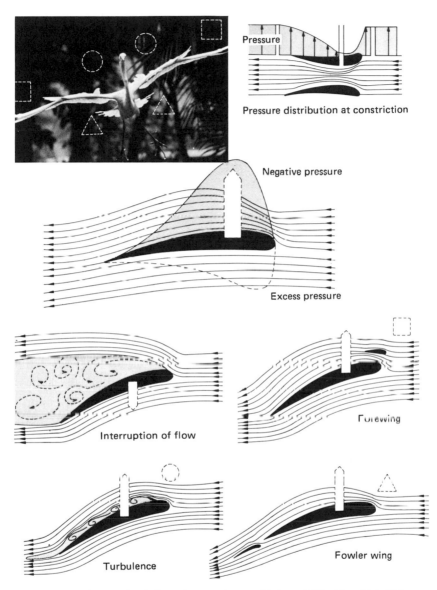

Pressure

Pressure distribution at constriction

Negative pressure

Excess pressure

Interruption of flow

Forewing

Turbulence

Fowler wing

For millions of years, birds have been using sophisticated lift aids which flight technology had yet to discover. Shown: a heron landing.

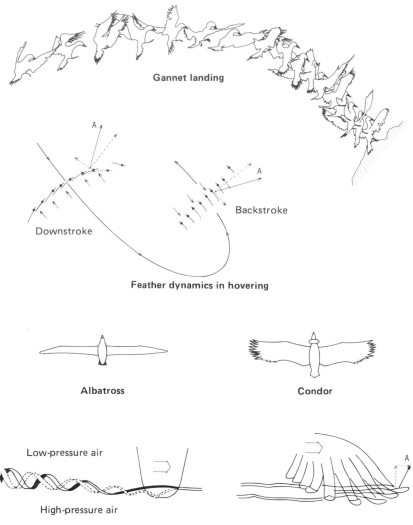

Gannet landing

Downstroke

A

Backstroke

A

Feather dynamics in hovering

Albatross

Condor

Low-pressure air

High-pressure air

A

Wingtip

Feather cascade

Many refinements of bird flight result from the complex dynamics of the individual feathers—among them the generation of lift during hovering and the prevention of spiral eddies at the feather cascade.

This black vulture, photographed exactly from the rear, shows the fanning out of the feathers at the wingtips.

Many interesting details about the flight of birds are insufficiently understood. One flying device that has received much attention is a group of feathers that spread out like a fan at the wingtips of slowly flying birds. They can easily be seen as working like forewings connected in series and thereby preventing the interruption of the air flow above the wing tips. Moreover, the numerous small winglets would add to lift and thrust. Obviously, however, the most important function of this feather cascade is the suppression of spiral vortices, which normally follow the wingtips and consume a lot of kinetic energy. These vortices are due to the pressure difference between the top and the bottom of the wing, which allows air to rise around the tip so that the air stream is diverted sideways and a rotating motion is imparted to the air. The photograph showing a black vulture sailing in an upwind gives an idea of the precision with which the feathers at the wing tips can be aligned.

Since no wing is equally well suited for every purpose, birds with different flight habits have different wings. Wings with a large surface generate very efficient lift, but have the disadvantage of sizable drag. Slow-flying birds like the eagle and the owl can take this disadvantage in stride. An important factor is the wingspan. Birds with large wingspan and narrow wings need less energy for flying than do birds with short and wide wings of the same wing surface, but on the average they need to fly faster since the air flow is more easily interrupted around narrow wings. This is why fast-flying birds such as the albatross and the swallow have developed long, narrow wings. Short, wide wings have been developed by slow-flying birds like the vulture and the oilbird. Wings often take part in a change of direction. For this purpose they can be pulled in on one side or flapped asymmetrically.

There is also a wide range of uses for the bird tail: as steering device, as brake, and as auxiliary wing. As with many other details of bird flight, well-founded biophysical research is still lacking.

What makes the flight of a bird such a fascinating spectacle is, no doubt, the harmonious and dynamic combination of ingenious physical technology and artistic beauty.

In the dense jungles of the Amazon an encounter with a hummingbird is impressive. In the brooding heat of the day the chorus of animal voices has fallen silent. The only audible sound is the quiet immersion of the paddle as your canoe slowly glides along a waterway roofed over by the jungle greenery. Suddenly, the sulky air carries a sound that is very distinct, though of low frequency. Before you can clearly establish the direction of its source, you feel that it is now coming from an entirely different direction. Perhaps it disappears for an instant, then comes back again. Your trained eye spots a hummingbird, only several centimeters long, hovering in front of a blossom. Its dark green plumage merges with the foliage. It stays only a few instants before plunging back into the jungle in a wild zigzag maneuver.

The many hummingbirds that enrich the Americas from Alaska to Patagonia with their colorful beauty have been called "flying gems." "Flying power plants" sounds less poetic, but it would be an appropriate term to describe their biophysical capacities. These tiny miracles of flight technology had to pay a high price for their acrobatic flying abilities. In order to keep their overbred flying apparatus running, they must ingest every day a quantity of nectar and insects that may exceed the few grams of their body weight. They spend the whole day searching for food, and enjoy only a short life span.

What are the aeronautical feats in exchange for which nature accepts such serious drawbacks? Hummingbirds beat their wings 18–78 times a second. They attain speeds of 80 kilometers per hour (*Archilochus colubris*). They can hover repeatedly and for long stretches. Their extremely flexible flight organs allow them to take off from a hover in any direction, and even to fly straight up from a resting place. An enormous set of flight muscles—up to 25 percent of the hummingbird's

The motion by which a hummingbird moves to another blossom. In hovering (center), it moves its wings in a horizontal figure eight so that only the vertical lift forces are maintained.

total weight—is available to power two very flexible wings. The wing arm of the hummingbird is remarkably short. The wing hand, whose inclination can be easily set by the bird in flight, constitutes the major part of the wing. But the unique feature of the hummingbird wing is that it is flat and resembles an elastic membrane. It can be flexed toward either of its two surfaces by a combined rotating and stretching movement. The alternating curvature of the wing in opposite directions is an essential ingredient of the hum. For this maneuver the humming-bird usually takes up a vertical position in the air while the wingtips trace a horizontal figure-eight. The inclination of the wing during this maneuver is always set so that lift is generated during both the down-stroke and the upstroke. Since each wing sweeps periodically one-third of a circle, the resulting lift effect can be compared to that of a helicopter. Lift is nothing but low pressure above the wing. The periodically oscillating hummingbird wing leaves in the plane of each stroke a pressure gradient which seeks equalization. Thus a current toward the ground is generated which corresponds to the lift. Since the two wings whirr in opposite directions, their torques are compensated—exactly as in an "eggbeater"-type helicopter.

Hummingbirds are miracles of flight technology, but at the same time victims of their own perfection. We humans ought not only to marvel at their charms, but to let them remind us of our own destiny: We too have in some respects become slaves to our technological genius and our material needs.

The cold Humboldt current, fed by antarctic waters, causes a vertical updrift at the coast of Peru and northern Chile which, with its high content of plankton, feeds huge swarms of "*anchovetas*" and other sardines, no more than a few centimeters long. These tiny, nimble swimmers are the diet of some 20 million guano birds, which hunt them on the open sea and sometimes several meters beneath the surface of the water. Cormorants, which are superbly equipped for underwater hunts, dive for them from the surface. Birds of less weight, like the tropical gannet (*Sula*) and the brown pelican (*Pelicanus occidentalis*), have learned to mobilize the necessary speed for diving by freefall from the air. This daring nosedive technique is not often used in nature. With a bit of luck, one can, in remote inland waters, observe ospreys attacking their prey in a nosedive and pulling it out of the water with their talons. The kingfisher too pounces headfirst on fish, starting from a low altitude. It is not rare for terns to try their luck with this technique. They keep continuous watch over the surface of the sea, sometimes hurrying along with rapid wing strokes, sometimes almost hovering. When they find a swarm of small fish, they hover several meters above, press their wings to the sides of their bodies, and let themselves fall into the water.

With its drama, perfection, and splendor, the collective nosedive attack of gannets on a school of fish is one of the most impressive natural spectacles. It often happens like lightning out of a clear sky. Numerous chains of rapidly flying gannets appear above an otherwise inconspicuous spot of the sea. They gather, quickly form groups, and suddenly start coming down on the water like lead, dropping from considerable height. They dot the surface of the sea with fountains of spray. For hours they keep the waters agitated in a merry-go-round of dive, hunt, take off, regroup, approach, and dive again. With the arrival of several relay teams of ponderous pelicans plunging headfirst into the water, flocks of cormorants ploughing through the waves, and clouds of gulls and terns flitting above the crests, the merciless drive reaches its climax. During this orgy of life and death, one can see the sardines jump out of the water in all directions. The sea seems to be boiling.

When I first witnessed such a hunt in the Galapagos Islands, I was so impressed that I decided to study more closely the nosedive maneuver of gannets and pelicans. But where on this wide ocean would

Gannets attacking a school of fish.

they be hunting tomorrow or the day after? Would I be able to muster the patience necessary for such an undertaking? I pursued the daredevil hunters between Ecuador and southern Chile. Perhaps half a dozen times I managed to get some stills and movie shots which helped me bring to light some of the physical details of their hunting technique.

The most daring nosedivers, the gannets, fish from the air exclusively. The approach to the dive occurs mostly at an altitude of 15–35 meters. The instant prey is sighted, a visible jolt goes through the body of the bird. The gannet brakes its flight and pulls in its wings slightly. Then it wings over to one side and turns, belly up, about 90° in a vertical position. With a few more flaps, it gradually bends its wings and systematically diminishes their surface. The wings ensure efficient steering up to the last moment through an intricate folding mechanism: the faster the bird moves, the smaller the wing surface becomes and the farther toward the tail end the stabilizing effect shifts. The gannet gradually and in correct technical sequence converts into a high-speed flying machine.

Depending on the altitude of the attack and on the wing effort during the first phase of the nosedive, gannets hit the water with speeds of 40–120 kilometers per hour. In the past, native fishermen disposed of them by nailing a sardine to a small plank floating under the surface. Diving gannets have been known to get caught in fishnets 15–20 meters deep in the water. In general, they remain underwater only briefly and immediately rise with hasty wing strokes to the next attack.

How does the gannet avoid veering off course and tumbling over during the dive and the dangerous moment of penetration? At 100 km/h, a slight gust, one wrong move, or rough seas could seal its fate. The secret was revealed by slow-motion photography: While diving, the gannet puts itself into a spin with a deliberate tail movement. The spin increases toward the point of impact as the bird lays back its wings like a figure skater bringing her arms close to her body. In a fast dive, this movement usually turns the body once or twice around its axis, acting like a gyroscopic stabilizer in a rocket. In the language of physics, the bird is kept on course by the conservation of angular momentum. This elegant mechanical stabilization notwithstanding, the moment of impact on the water surface is critical because of the powerful forces involved. But the gannet has been well primed by nature for this moment. Its body can stretch into an ideally streamlined spindle. Any unevenness about the head is eliminated. At the moment of immersion, the gannet draws in its neck slightly so that the pointed beak and the flat top of the head form a straight continuous line with

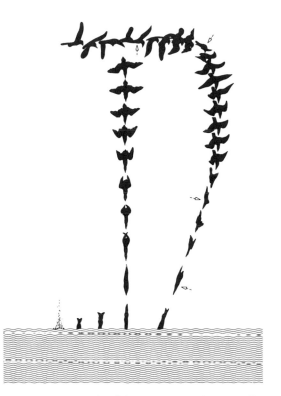

The nosedive attack of the gannet. In this overall view of the maneuver (picture intervals: 1/12 second), the autorotation of the bird is discernible (arrow, belly side). The vertical picture sequence shows the aerodynamically sophisticated folding of the wings before impact (picture intervals: 1/24 second).

the body, creating a cone which combines low resistance with high stability. The maze of air cells between the skin and the muscles, directly or indirectly connected with the lungs, receive and distribute whatever pressures occur.

Nosedive hunting in groups requires strict discipline. Small units of gannets go into a dive simultaneously and in parallel, as if heeding an invisible signal. If the group is larger, the shifts are staggered. A front line of diving birds rolls like a carpet forward or backward to the direction of approach. After plunging into the water, all the members of the unit duck in the same direction to get out of the way of the subsequent divers. In this way they can even influence the escape direction of the fish. In the tumult of the big fishing expedition there appear stationary attack and departure routes into which even newly arriving units of birds can be channeled. Accidents are very rare; evolution must have mercilessly eliminated all traffic violators.

The nosedive maneuver of the pelican is both powerful and amusing. From a height of several meters up to 20, the slowly flying bird goes into a sideward slip and plunges steeply or in a slight spiral toward the water, its wings slightly bent and its head drawn back. With neck stretched way forward, it disappears into the water with a loud plop. Popping up right away like a cork, it goes about devouring with gusto the fish caught in its pouch. If conditions are favorable it continues the hunt right from the surface of the water.

Nature's most skillful gliders, the albatrosses and the large fulmars, have made the open sea their home. Some members of this elite club—the royal albatross (*Diomeda epomophora*), the wandering albatross (*Diomeda exulans*), and the southern giant fulmar (*Macronectes giganteus*)— spend their lives on the icy air currents over the antarctic sea and touch land only when they brood. Surely they circle the globe many times. The very narrow wings of large albatrosses can be up to 3.5 meters long and are extraordinarily specialized for effortless sailing in the wind.

Albatrosses seldom flap their wings. They often must make a considerable effort when lifting off from the surface of the water. On land, they do not shy away from a lengthy walk in search of a windy cliff from which to take off. The wind is an altogether vital element. The average wind velocity on Campbell Island, one of the most important albatross nesting sites, is 50 kilometers per hour. The first time I observed albatrosses in flight, I was struck by their elegance and beauty and altogether enchanted with the serenity and ease these giants radiate when they abandon for a while their mysterious routes to escort a ship.

One should suppose, in theory, that headwinds would be a drawback for a gliding bird and would soon exhaust its strength. Just the opposite is true for the albatross, which has come up with a sophisticated technique to circumvent and exploit the laws of physics. The details are not quite clear yet, but it is certain that the albatross makes use, at least in part, of the aerodynamic lift generated by the friction between air and waves. This alone, however, does not explain its magnificent performance. Wind velocity is at its lowest at sea level (because of friction), and increases gradually to an altitude of about 12–15 meters. The albatross takes advantage of this physical fact by gaining enough energy for its flight from the moving air masses. An albatross glides from an altitude of about 10–15 meters down to the water at an angle against the direction of the wind, attaining considerable speed. Just above the water it turns sharply against the wind.

Brown pelicans in a collective nosedive (coast of central Chile).

The albatross is able to fly over the sea for hours without upwinds and without ever flapping a wing.

Regaining altitude quickly, it slows down correspondingly. Since wind velocity increases with altitude, the relative speeds of albatross and wind remain by and large unchanged. Before the bird slows down too much, it again wings over and glides downward. Under favorable wind conditions, the albatross need not beat its wings a single time; it manages to recover from the air currents whatever energy was lost to friction.

To understand this astounding feat to some extent, imagine a rotating tennis ball "suspended" under a slanted air jet. The autorotation of the ball causes the air flow above to be faster than that below, where the ball rotates against the current. The lift generated in this manner compensates for gravity. Similar flow forces appear when a tennis ball is "chopped" so that it flies in a curved path. The appearance of forces under such conditions of uneven flow around a rotating object is called the Magnus effect.

The albatross may use the same principle of lift generation. Of course, it does not need to rotate, since the air current increases toward the upper side of the wing anyhow. The bird can soar in any direction, if the wind is favorable, by choosing the proper direction of descent. If, however, it chooses to continue its flight without flapping its wings, it must retain a certain minimum speed when turning against the wind. This speed must be greater the weaker the wind (more precisely, the less the wind increases with altitude). This is how the albatross can also fly against the wind: the stronger the wind, the better the soaring conditions close to sea level. Albatrosses have been observed soaring at wind velocities of more than 100 km/h.

When sailing close above the water, the albatross also makes very good use of the kinetic energy of the waves. In this it is matched by its cousin, the southern giant fulmar (*Macronectes*), which has a wingspan of up to 2 meters. I observed two of these narrow-winged birds of prey for hours as they were following—and without a single wing stroke—a boat traveling through the Straits of Magellan at the southwestern tip of Chile. The technique they employed was simple and ingenious. A boat leaves a wedge-shaped wake. The fulmars approached from the outside between two waves and placed themselves directly in front of a crest, no more than a centimeter or two above the water. The velocity component of the wave in the flight direction accelerated the birds, which controlled the position of their wings with precision, and propelled them toward the inner edge of the wave and the turbulent zone of the boat. Then the birds pulled above the turbulence in a steep arc, transforming their speed energy into potential altitude energy. From this height they could then return in an easy

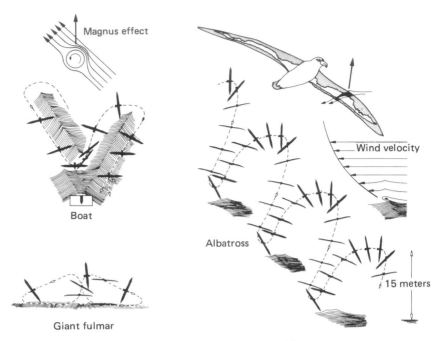

The flight movements of the albatross and the giant fulmar.

glide and again, in a wide arc, approach a trough from the outside. They regularly alternated the sides of the approach. That this flight pattern is designed for the artificially produced wave pattern of a ship, and does not occur in nature, proves the astonishing adaptability of these gliding birds.

Albatrosses and fulmars need precise information about air currents for their acrobatics. Their nostrils end in horny tubes (they belong to the order of the tube-nosed swimmers, *Procellariformes*), and it is possible that these tubes are sensitive organs for measuring currents, like the static pressure tubes of human technology.

The flight of insects

Dragonflies are able, without any major effort, to accelerate in a very short time from a hover to about 15 meters per second. This, at least, is the impression conveyed to the observer of their fascinating chasing games. They flap their wings at a frequency of about 20–40 strokes per second, which for insects is not very high. Bees and houseflies manage some 200 strokes per second. Some mosquitoes (for instance *Culex*) execute 600 strokes, and the tiny gnat of the genus *Forcipmyia* 1,000, per second.

This giant fulmar has placed itself centimeter-close to the updraft of a traveling wave crest, which propels it forward above the water without any effort on the bird's part.

Impressed by such feats, one might conclude that insects have above average strength in addition to a flying apparatus of extraordinary aerodynamic design. But close inspection of their wings brings a surprise: There is no streamline construction at all. The cross-section of a dragonfly wing shows an angular, plicated structure. The wings of many other insects look like corrugated plate. Since evidently nature succeeded masterfully in solving the flight problems of insects, these inventions must be in accordance with the laws of physics. And so they are. By taking a closer look at these matters, we can add a thing or two to our knowledge of physics.

The physical factor mainly responsible for the difference between the flight technique of birds and that of insects is size. Whereas muscle strength is proportional to the cross-section of the muscle (that is, it decreases with a decrease in surface of the animal), weight drops in proportion to the animal's volume. An insect, therefore, has exceedingly strong muscles for a very light body. The generation of great mechanical forces is further facilitated by the fact that oxygen can very quickly be replenished through the surface, which is large relative to the volume of the insect. Only on the basis of these premises is an insect able to keep up the high frequency of its wing movements. Insects could fly much faster than birds if not for the friction drag, which drops at a much slower rate than body weight. Insects need to invest a much larger share of their energy to overcome air friction than do birds.

Of particular importance in the development of the flight devices typical for insects was the fact that the dimensions of a flying body also affect the laws of fluid mechanics. Given the same density and viscosity of the flowing medium (air), the Reynolds number escalates with the speed of the flying body and with its length. The flow around two objects of different size is equal only if the Reynolds number is the same. Since insects are small, their Reynolds numbers are relatively low, notwithstanding their high speed and wing-stroke frequency. The Reynolds number is around 10,000 for large dragonflies and much lower for smaller insects. At such low Reynolds numbers, flows are characterized by a very interesting feature: though it is necessary to maintain as even and laminar a flow as possible when the Reynolds numbers are higher, quite the reverse is the case for lower numbers—turbulence is an advantage. Experiments have shown that in such cases turbulence increases the lift and at the same time greatly lowers the resistance offered to the flying body. A smooth, aerodynamic wing profile for insects would, therefore, have been not only superfluous but actually disadvantageous. The optimal wing for an insect

The pleated wings of the dragonfly are crisscrossed by numerous supporting veins.

The wings of insects are not streamlined, but are unevenly constructed, with wavy or pleated surfaces.

is a simple curved or curvable plane with turbulence-producing irregularities.

The uneven wing has been used by nature to the great benefit of insects. Insect wings are unsurpassed masterpieces of lightweight construction. One square meter of a dragonfly wing would weigh only about 10 grams, many times less than the finest paper. With all that, the extraordinary stiffness and sturdiness of the dragonfly wing is due only to the unevenness of its structure. The front part is folded in zigzag fashion, with reinforcements parallel to the front edge of the wing located in the corners. The hind part, on the other hand, is elastic, composed of many hundreds of polygons. The wavelike, uneven wing surfaces of other insects serve a double purpose: They guarantee dynamic stability against stress, and they produce turbulence. They often have additional features that serve as camouflage.

The flying motions of insects resemble in some respects those of hummingbirds. Since insect wings are reinforced mainly at the front edge, whereas the hind part is elastically malleable, they passively flex in either direction during the wing stroke. Good flyers move their wings dynamically without interruption in the shape of a figure eight. During the downstroke from top rear to bottom front the wings are held in a relatively horizontal position so that mainly lift is generated;

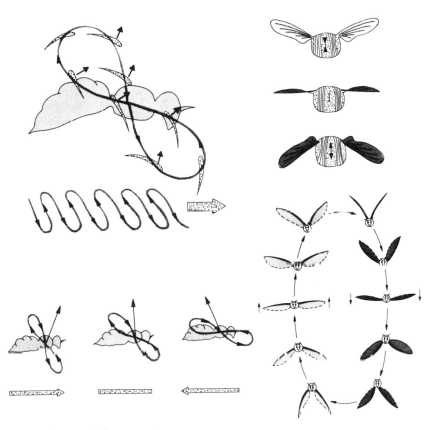

Wasps, bees, and flies owe their aerobatic abilities to highly efficient flight organs and to their largely automatic high-frequency movements and wing settings.

during the back stroke the wing is moved almost vertically so that the main result is thrust.

Hoverflies, hawkmoths, and certain hymenoptera are able to hover in place just like hummingbirds. They can even fly backward. They execute these maneuvers simply by changing the inclination of the figure eight traced by their wings, whereby they also modify the angle of inclination of the resulting flight power (see figure). In contrast to birds, insects are not able to vary the plane of their wings, nor the stroke frequency. On the other hand, being light and unstable, they need efficient flight-control mechanisms. As far as we know today, steering in flight is effected mainly by one-sided stronger or weaker wing strokes, or by changing the wing's angle of incidence. Only the dragonfly has flight muscles that connect directly with the wing root. All other insects use indirect driving mechanisms. The flight muscles are located in the chest and cause a periodical bulging and flattening of the dorsal plates. This movement is transmitted mechanically to the wings, which rise and fall in the same rhythm. No nerve could be quick enough to beat the time for this rhythmic movement of 1,000 strokes per second. Rather, the natural mechanical frequency of the chest and of the attached flight apparatus is stimulated when the insect flies. To date we have only some vague theories regarding the bio-energetical foundations of these oscillating flight systems.

Flying seed

While I was traversing the hot wilderness of the Gran Chaco in the northwest of Paraguay, the truck on which I had hitched a ride broke down in an uninhabited prairielike region. We were in for a rather long wait. While the others reacted rather sulkily to this forced rest, I came upon a surprisingly interesting and amusing plaything: flying seed.

After I had come to know several kinds of flying seed, I began to study and to compare their various flight properties. Seeds of the big-bellied thorny bottletree were sent off on their journey wrapped in loose round cotton pellets. This gave them a very large surface to offer to the wind and served as a brake in the descent to the ground. At the same time the seed pellets were very light, since they consisted almost entirely of encapsuled air. Moreover, when falling upon bare sandy ground, they would roll before the wind. Other seeds I noticed were imbedded in a very thin single wing. These allwing gliders had a kind of indentation in front and showed a very odd flight behavior: If sent sailing at a downward angle, they would pull upward after a short stretch of acceleration, then turn over and continue their glide upside-down in the opposite direction. Repeating this maneuver many times, they would tumble down very slowly but vertically in a zigzag motion. This acrobatic flying maneuver evidently serves to slow the descent of the seed. I discovered two advantages of the periodical pulling up and turning over of the seed: Kinetic energy was again transformed into potential altitude energy, and the somersault presented the broadside of the wing to the wind. A wobbly flying motion seems sometimes to be more advantageous than smooth gliding.

The two kinds of seed described above typify the structural principles of most flying seeds. In one group, the seed is as light as possible but is equipped with a very large surface which facilitates hovering and drifting in the wind. The parachute seed of dandelion belongs in this group. The lightweight technique is carried almost to the extreme; orchid seeds often weigh no more than a millionth of a gram. In the second group of flying seeds, sophisticated lift and flight mechanisms

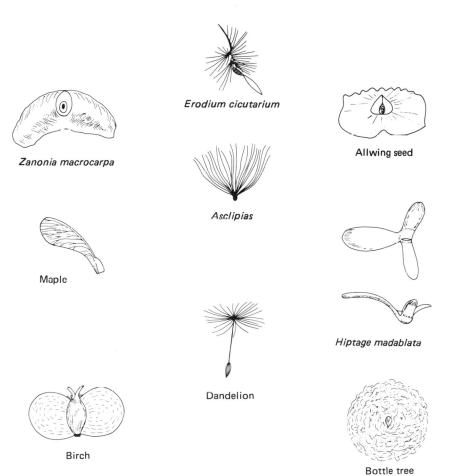

Erodium cicutarium

Zanonia macrocarpa

Allwing seed

Asclipias

Maple

Hiptage madablata

Dandelion

Birch

Bottle tree

Some of the imaginative flight apparatuses of the flying seeds.

have been developed. In this group belong all seeds with gliding surfaces or wings. The allwing-glider seed of the tropical palm *Zanonia macrocarpa*, some 15 centimeters long, has become well known as study object of flight pioneers. This seed is an astonishingly stable glider which descends slowly and in wide circles to the ground.

Other seeds and fruits use natural rotation to generate lift for braking. The asymmetrically winged seeds of the maple spiral their way down to earth in screwlike motions, whereby the wing rotates like a propeller, generating lift and slowing the descent. Since the seed starts rotating all by itself, its energetics and dynamics can be compared to those of a windmill, which derives energy from air currents. In the descent of the seed this energy functions as a brake on the self-movement. The flying fruit of the birch, only a few millimeters in size, could easily be taken for tiny insects. Not only do they sport well-formed wings on both sides of the oblong body, they also have two distinct antennalike appendages. When I first noticed them, I thought that tiny flies had infested my place of work. I wonder whether they do not become stowaways in insect hunting birds by reason of their amazingly genuine appearance. This would be quite an astonishing example of mimicry. Nature does come up with an extraordinary variety of ideas, even if it is just a matter of sending tiny seeds on a journey with the wind.

Flight experiments

The pterosaurs of prehistoric times are known as the mightiest fliers ever produced by nature. Fossil specimens of the genus *Pteranodon*, with a wingspan of 6–8 meters, have been known for quite some time. The recent discovery of the remains of three more pterosaurs in Big Bend National Park in Texas astonished the scientific community. When the well-preserved huge bones of the wing arm were used to calculate the wingspan on the basis of the design of heretofore discovered pterosaurs, the result showed a value of 15.5 meters. Even if the wings of giant saurians had measured no more than 10 meters, they would present a phenomenon difficult to understand from the aeronautical point of view. It has been calculated that the heaviest animal that could still lift itself from the ground by its own strength could not exceed 16–20 kilograms. This weight limit results from the fact that muscle strength increases with the cross-section of the muscle, while weight increases with the volume of the muscle. It is precisely by reason of this law that man will never be able to fly by his own strength, inasfar as he takes bird flight as his model. However, by choosing extremely light, rigid wings and transmitting our body strength mechanically to a propeller, we create more favorable physical conditions which technology can cope with, as evidenced by the crossing of the English Channel in the pedal-driven *Gossamer Albatross.*

Since the bones of the pterosaurs were thin-walled and hollow, we can assume that these creatures were of very lightweight build. The fine elastic flying membranes which stretched between forelimbs and body were certainly also very light. It is therefore still possible to visualize as aeronautically viable a pterosaur 6–8 meters in length and weighing 15 kilograms. But such a creature certainly was no aerial acrobat in its main occupation, fishing along the seashore. It has been established that the recently discovered giant pterosaurs lived far inland. They may have lived the lazy life of scavengers—possibly by using continental upwinds, like the largest land bird existing today, the condor.

I had many occasions to watch condors in the southern Andes. With a wingspread of 3.5 meters and a body weight that can exceed 12 kilograms, the condor moves within critical limits as far as flight is concerned. Like the captain of a sailing vessel, it must sometimes wait for favorable winds when all is calm, or cast a worried glance at the load it carries. Usually the condor leaves its dung bleached lookout post high up on the rocks only toward noon, when the warming rays of the sun have put the machinery of the upwinds in motion. It is seldom seen to actively flap its wings on its slowly circling reconnaissance flights or fast raids across the mountain landscape. Mostly the wings are used only to traverse an adverse air current. The people of the Andes used to outsmart and catch the condor by taking advantage of its aeronautical weakness. They would surround a carcass with a low fence of timber which the condor could easily fly over. After the condor had gorged itself on the carcass, the short runway inside the enclosure would not be adequate to permit a takeoff, and the bird would be helpless.

The example of the condor hints at the reason why the much mightier prehistoric pterosaurs eventually became extinct. In the course of evolution, nature has again and again experimented with various flight mechanisms. Fishes, mammals, and insects, besides reptiles and birds, have managed to penetrate into the airspace. The laws of aerodynamics often fashioned their wings and gliding surfaces in remarkably similar ways, but, just as in the history of the airplane, extreme and arbitrary ideas surfaced occasionally. An interesting example are the flying fishes, whose aerobatics astonish travelers in tropical seas.

Flying fishes (*Exocetidae*), of which there exist some three dozen species, escape the eyes of their enemies by emerging into the air. Their wings are formed by pectoral fins fanning out into huge gliding surfaces. They are too fragile and equipped with too few muscles for active flying movements. But nature had resort to another kind of driving mechanism. The flying fish accelerates its body for the start and lifts it with spread-out wings so high above the water surface that only the reinforced lower part of its crescent-shaped tail fin remains immersed in water. In this position it flips the fin 50–70 times per second and thereby produces the necessary final velocity for a sailing start on a runway 10–20 meters long. While riding with spread-out wings on the submerged and paddling tail fin, the fish operates exactly according to the energy-saving principle of a hovercraft which is lifted high above the water and driven by a submerged propeller. As soon as the "hovering" fish has reached a takeoff speed of 50–75 kilometers per hour, it lifts the tail fin out of the water and sails for about 50 meters like a toy kite. Then it touches down on the water, tail fin first, and is able to accelerate without further immersion for the next takeoff by using its "flyingboat" drive. It is not rare for a fish to cover 400 meters at one combined acceleration and gliding stretch. The active flapping of wings during flight has been established only for one family (*Gasteropelecidae*), for which the technical prerequisites are most favorable. As is the case with many birds, the weight of their flight muscles comes to one-fourth of the total body weight. These true flying fishes have a correspondingly well-developed chest, on which they ride like on a runner while their winglike pectoral fins flutter above the water and produce a whirring sound. But it is rare for these strange fishes to completely lift themselves above water for as much as several meters. They prefer to plough through the waves like a flying boat with air propulsion.

Among mammals, too, nature has brought forth some powerful flyers. Bats (*Chiroptera*), among them both the fruit bats (*Megachiroptera*) and the insectivorous bats (*Microchiroptera*), have elastic flying membranes stretched between the greatly extended bones of their fingers and hands and reaching backward as far as the hind legs and the tail. The largest representative of these nocturnal flutterers, the red-necked fruit bat *Pteropus vampyrus* of southeast Asia, attains a wingspread of 1.5 meters with a body length of 40 centimeters. Compared with birds, the flight of bats seems inelegant. But hardly a bird can compete in agility with these mammals. They have developed a method of flying that is ideal for their nocturnal world with all its obstacles. There is an almost continuous spectrum of specialized flying techniques, extending from the oscillation in place of those species that feed on flowers to the gliding ventures of fruit bats. The wing stroke of bats is more sophisticated than it appears to be. The wingtips trace an

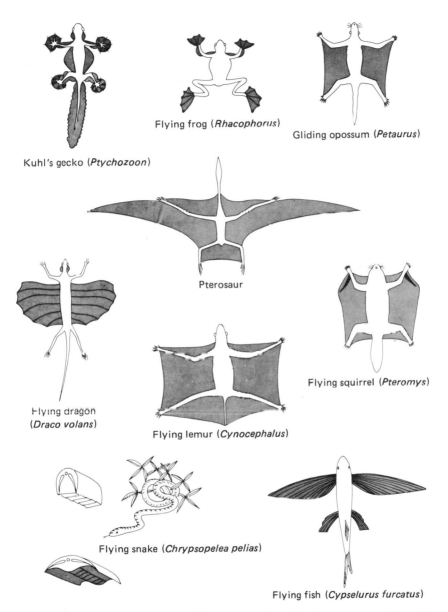

Kuhl's gecko (*Ptychozoon*)

Flying frog (*Rhacophorus*)

Gliding opossum (*Petaurus*)

Pterosaur

Flying dragon
(*Draco volans*)

Flying lemur (*Cynocephalus*)

Flying squirrel (*Pteromys*)

Flying snake (*Chrypsopelea pelias*)

Flying fish (*Cypselurus furcatus*)

Nature has again and again sent flight pioneers into the air to test new ideas. But except for the pterosaurs, whose flying properties are difficult to reconstruct, the animals shown here have achieved only an awkward form of gliding.

elliptic path. The wings are wide open when they beat downward, and are almost parallel to the body, folded forward and with the narrow side up, when they are pulled back. The smaller the bat, the faster it flaps its wings—the small horseshoe bat flaps 15–20 times each second. Rapidly flying bat species have long narrow wings, just like rapidly flying birds. The long-winged bat (*Miniopterus schreibersi*) and the common noctule bat (*Nyctalus noctula*) attain velocities of 50 kilometers per hour.

Neither have ideas for gliding been lacking in the animal kingdom. At least six species of mammals—the pygmy flying phalanger (*Acrobates*), the gliding possum (*Petaurus*), the greater gliding possum (*Schoinabates*), the scaly-tailed squirrel (*Anomalurus*), the flying squirrel (*Glaucomys, Petaurista*), and the flying lemur (*Cynocephalidae*)—have developed large gliding membranes for this purpose, which are stretched between the forelimbs and the hindlimbs. For the flying lemurs, the tail too is enclosed by the membrane, while for the rest of the gliders the long tail is an independent steering device. The common giant flying squirrel (*Petaurista petaurista*) from southern Asia, which glides on a 60-centimeter-wide flying membrane, bridges gaps of 60 meters between trees. Flying lemurs jump as far as 70 meters. Many of these hairy sailors are even able to change direction in flight by steering with the tail and by altering the tension of the membrane.

Reptiles too were repeatedly enrolled by nature as flight pioneers. The flying frog (*Rhacophorus*) rides on large winglike flaps stretched between its toes. In the flying dragon (*Draco volans*) the first five or six ribs have been transformed to support a semicircular parachute which allows the lizard to bridge 10-meter gaps between trees by gliding. Truly amazing is the gliding technique of the south Asian flying snake (*Chrypsopelea ornata*). When getting ready to jump, it extends its body and pulls in its belly so as to form an indented trough. In this position it manages to make a controlled descent by riding on a well-stabilized air cushion.

Who knows how lavish an assortment of extravagant designs for flight have been tried out throughout the millennia and have in the end been discarded on the garbage dump of evolution!

4

Acoustic Techniques

Some of the nights I spent in tropical wilderness areas were unforgettable acoustic experiences. In a camp or a hammock in the jungle, a feeling of menace surrounds you like a black curtain just a few feet away and keeps your ear on the alert. Each unfamiliar sound, each change in the rhythm of the night music is carefully registered when you cannot sleep. If you travel a lot in the Amazon region, you learn that the timbres of the night can vary a great deal. Often the tone is set by millions of cicadas with their shrill song, accompanied by jungle frogs with their dull, loud, rhythmic "drum beats" and a motley orchestra of colorful voices. Then again there is dead silence, broken only occasionally by the almost inaudible sounds of bats on the hunt. The silence is torn by the shriek of a night bird, which a chorus of hundreds of alerted animals takes up until it slowly abates and dies away. Again and again the bird repeats its call and the chorus follows. And then suddenly the death cry of an animal interrupts the monotony and sets off a cacophony of all the voices. It does not take long, however, till a rhythmic beat like a lullaby again proclaims peace. On such nights I sometimes thought with wonder about nature's having made sound so indispensable a means of communication between creatures and such a sharp tool for hunting and navigation.

From the viewpoint of physics the generation of sound is a very simple mechanism. Any object that is touched and set in vibration produces sound. With each forward vibration it presses the air or fluid particles away, and with each movement backward it leaves an empty space into which they can flow back. The particles pass the thrusts they receive on to their neighbors, so that periodic waves of increasing and decreasing density propagate from the vibrating object. These waves of periodic pressure fluctuations are the sound waves. They contain energy, and can therefore transmit signals if a suitable receptor catches them. The sounds we hear depend on the frequency with which the pressure fluctuations reach our ears. When we hear a middle C, our eardrums are being struck each second by 256 waves of higher pressure and as many of lower pressure. Since sound travels through air with a speed of 330 meters per second, the distance between two wave crests in air is 130 centimeters. If the same sound is produced in water, the zones of higher pressure move at intervals of 550 centimeters, because sound travels faster in water. Water cannot be compressed, and can transmit pressure thrusts more easily. It conducts sound a distance of almost 1,500 meters per second.

Like light waves, electromagnetic waves, or water waves, sound waves can be reflected, refracted, or absorbed. Echoes are caused by reverberating sound. The whispering galleries of the Middle Ages have elliptical walls which reflect sound in such a way that a word whispered at one focus of the ellipse is heard without effort at the other focus, but nowhere else. Walls need not be mirror-smooth to reverberate sound. Whatever unevenness there is need only be much smaller than the length of the reflected wave, and the sound waves we can hear are several hundred thousand times longer than light waves. Just as with light, an obstacle placed in the path of sound produces a "shadow" that is free of waves. If the obstacle is smaller than the sound waves, the waves are dispersed in all directions. Like water or light waves, sound waves can cancel or reinforce each other when they overlap. They can also be bent around corners, since new circular waves can radiate from each point where air or water particles are set in vibration. (This is much more easily achieved with low-frequency sound than with high-frequency sound.)

The laws for sound are similar to those for electromagnetic waves. Good television reception requires waves beamed directly from the station, whereas medium-frequency waves can still be well received from behind an obstacle. A waterfall sounds dull from behind a rock, but its roar makes a very high sound when you step out from behind the rock. Sound slowly loses energy when it travels through air, water, or solid objects. This energy is converted into heat—that is, non-directional kinetic energy of molecules. Sound absorption increases with the square of its frequency: deep tones are heard considerably farther than high tones.

The laws of acoustics were already in existence when life was still in its most primitive state. At the current stage of evolution, these laws are put into action with all available sophistication for life's benefit.

How animals prooduce sounds

Any vibratile body can be used as a source of sound, be it a string or membrane that is put under stress from the outside, a rigid three-dimensional body with elastic properties, or an air column inside a suitable device. Musical instruments are outstanding examples of all kinds of mechanisms for sound production. But not every vibrating body is necessarily a good source of sound. Often a body is too small, or its vibrations are not wide enough, to impart an effective push to the surrounding air. For this reason, the sound generator often is connected to a second vibrating system of larger surface or volume. This sound box, or resonator, is usually dimensioned so as to vibrate with the same frequency as the sound generator, in which case the energy transmission is especially efficient. This resonator principle, which is also the basis for the sound boxes of string instruments, is quite frequently utilized in nature.

The sound organs of animals often have surprisingly simple physical mechanisms. Insects generally communicate by means of "string instruments." In most cases a hard chitin edge is rubbed against a comblike stridulation ledge to produce the sound. Sounds can be produced by rubbing together various parts of the body: wing against wing, leg against wing, leg against abdomen, head against thorax, or thoracic segments against each other. Crickets, for instance, rub one edge each of their two forewings against each other.

Many insects produce sounds that reach far into the region of ultrasound. Locusts attain frequencies of 90,000 hertz. Sound waves of such frequencies are only about 3 millimeters long. Insects have no problems in producing such frequencies, since their sound instruments are of similar dimensions as the wavelengths of ultrasound. The sound generated by these insects is of a very complex structure. Every time a tooth of the "comb" hooks in, a chain of vibrations is produced which slowly fade until the next tooth connects. Another simple principle of sound production is used by leafhoppers. They have an elastic chitin plate on the underside of the abdomen. The plate is reinforced by ribs, which they indent periodically by a muscular movement. The bass, the freshwater drum, and the sea robin use muscles to drum on their air bladder. Several kinds of mackerel scratch with their grooved teeth like a stylus on a record. Other marine creatures rub fins or parts of their skeletons together. The rattlesnake's rattle is produced by hard, loose scales at the tip of the tail.

Most mammals and many amphibians have highly developed wind instruments in their throats. The larynx of mammals (including man) is made narrow by two skin folds—the vocal cords—with only a small opening (the rima glottidis) between them. When air is exhaled through the throat, the vocal cords vibrate and generate sound. The shorter the vocal cords are adjusted and the tauter they are stretched, the higher the sound. The oral and pharyngeal cavities (which can easily change shape) and the lungs (which can be inflated or compressed) constitute the sound box that amplifies the vibrations of the vocal cords. The sounds may be further amplified by additional resonance sacs in the area of the larynx, such as those possessed by howling monkeys.

Bats generate their ultrasounds by the vibrations of the finest of skin folds in the larynx. Shrews likewise emit ultrasound from the throat. Small fruit bats (*Rousettus*) generate sound by flipping their tongue. Frogs and toads set the margins of their larynx in vibration by pushing air jets out of their lungs, making the slitlike opening of the larynx open and close alternately. Mostly it is the inflated throat which serves as resonance body for the amplification of the sound, but some species have additional resonance sacs which enlarge the mouth cavity. "Golden throats" are also found among birds. Their voice organs, however, are located lower down, where the windpipe merges into the two bronchial tubes. Here they have tympanic membranes which can be made to vibrate by air currents just like vocal cords. Songbirds are able to regulate their tones with numerous muscles. Storks, pelicans, and other less musical birds must forgo this privilege.

In the water, vibrating membranes and cords would be quickly muffled. This certainly is a physical reason why "wind instruments" have not caught on among marine creatures. Underwater, nature limits itself to drum and tympani sounds and crackling, scratching, and rattling noises. How do creatures which can produce nothing but crackling or scratching sounds manage to yell at each other? Fishes have found a way: they take up positions at close range and thrash about with their tail fins to produce low-frequency sound waves.

Small creatures generally produce higher sounds than big ones because their sound apparatus has smaller dimensions. The fact that human language happens to generate sounds between 100 and 10,000 hertz is a result of the dimensions of our speech organs. If we were tiny, we would communicate by ultrasound.

The mechanisms of hearing

In order to record sound, we can either measure the periodically varying speed of molecules or the pressure they exert on an obstacle.

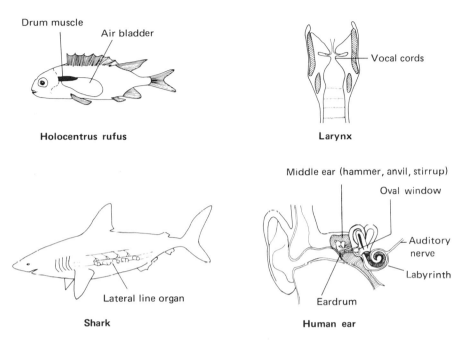

Nature has experimented with a variety of physical possibilities to create and register sound.

Almost all technological sensing devices register pressure fluctuations; they have thin membranes which are set in vibration by the sound and activate a suitable recording device. Most biological hearing mechanisms operate according to the same principle. In its simplest form it shows up in insects. Crickets, cicadas, and moths are equipped with large air-filled bubbles, whose walls serve as tympanic membranes to transmit vibrations directly or indirectly to the sensory cells. The "drum skins" of crickets are not located in the head, but in the lower legs. The hearing organs of locusts are on the thorax. Cicadas hear with their abdomen. Many moths are especially sensitive to ultrasound, and react to it by wildly trying to escape or by playing dead. Ultrasound to them is the "station identification" of their arch-enemy, the bat. They are hardly able, however, to distinguish various frequencies; whatever generates sound above a certain frequency they interpret as a bat.

In order to enable humans and other vertebrates to distinguish clearly between various frequencies, nature had to cope with tremendous acoustic problems. Sound composed of many vibrations had to be disassembled into its components so that each component could be recorded separately. This feat was successfully accomplished in a saclike fluid-filled organ, the cochlea, which in man is spiral-shaped. In order for sound to be funneled with little loss of energy into the cochlea, the acoustic resistance of the environment and that of the inner ear had first to be coordinated. Let us take the human ear as a model for a closer study of the device by which this problem was solved.

Our ear consists basically of three segments, each with its own task. The auricle with its external auditory meatus is responsible for catching the sound out of the air, focusing it, and setting a thin membrane—the eardrum—in vibration. Behind the eardrum is the middle ear, where the acoustic resistance (density × speed of sound) of air is equalized with that of the fluid as a prerequisite for the transmission of energy. At the entrance of the labyrinth is an oval "window" into which the sound is funneled. Then it travels, through the fluid, deeper into the cochlea, where it is received by nerve cells along the basilar membrane. The eardrum and the oval window are linked mechanically by three small bones (ossicles) named for their shapes malleus (hammer), incus (anvil), and stapes (stirrup). According to the laws of leverage, the ossicles transmit the vibrations of the tympanic membrane to the membrane stretched across the oval window. The surface of the eardrum is about 20 times that of the oval window. The pressure transmitted to the oval window from the middle-ear ossicles is proportionally increased when it reaches the window membrane. A further contribution is made by the well-synchronized mechanical transmission ratio of the bone levers: A strong deflection of the eardrum appears on the membrane of the oval window as a weak deflection, though it is the result of high pressure. Thus the sound can enter the fluid of the inner ear without any effort; the wide, low-pressure vibrations of sound in air have been successfully transformed into the short, high-pressure vibrations of sound in water.

We still do not know exactly what happens inside the cochlea, where sound is transformed into nerve impulses. It is considered certain that sound is separated into traveling waves of varied frequencies (von Bèkèsy) inside the narrowing cochlea, which is constructed so that sounds of different frequencies travel in it at different speeds and also fade or increase at different spots. Different wave forms are created in the basilar membrane according to the different pressure waves, and thus the composition of frequencies can be sensed by nerve fibers. Had nature not discovered the principle of the acoustic traveling wave, man would look quite different. Only the principle of resonance would have been available for separating sound according to frequency, and whole batteries of frequency-specific vibratile elements would have been required. One need only think of the size of a piano, which is tuned to the sound frequencies of human hearing between 18 and 18,000 hertz, in order to recognize that the principle of resonance would not have offered an alternative.

For a long time it remained unexplained how the human ear is able to establish the direction from which a sound is coming. Depending on the angle at which sound approaches, it reaches the two ears at a transitional interval of only one hundred-thousandth to several ten-thousandths of a second. No nerve can measure such short differences in time. Moreover, man can also distinguish between sound waves from above and from below, which reach both ears simultaneously and do not differ in the direction of their vibration. Today we know that our ears can distinguish between the direction-dependent sound patterns produced by sound dispersion at the auricle, the head, and the body.

Fishes live in an entirely different world of sound than terrestrial animals. They are in particular dependent on sensitive perception of pressure waves, without which they would hardly be able to find their way at night, in murky waters, or in great depths. Physics has imposed different limits on hearing mechanisms in the air and in the water. In this respect, too, nature has made masterful use of all possibilities. The lateral line organ of fish is an ingeniously simple and efficient device for the detection of sound and pressure waves. It is common to all fishes, though there are some structural variations. In a typical case, as for instance the shark, one tubular duct runs subcutaneously along each side of the body from head to tail. It is linked to the outside by membrane-sealed "bull's eyes." Pressure waves that penetrate into this fluid-filled duct system cause currents, which vibrate sensory hairs. The fish establishes the direction of the sound impulses from their distribution. A fish swimming past an obstacle gets a good idea of the environment through pressure reflection. Combined with the generation of sound through rhythmic tail movement, the lateral line organ could even serve as a primitive sonar device. In the head area, primitive hearing centers developed out of the lateral line organ. They function in that a body set in vibration by pressure fluctuations stimulates a pad of sensory hair. Most fishes hear only very low frequencies. For species with good hearing (silurids, minnows), the upper hearing limit lies at 6,000–8,000 hertz.

The marine creatures with the best developed and most flexible acoustic sense are dolphins and whales, which brought their skills from the dry land where their ancestors once dwelled. Dolphins have an intricate ultrasound navigational system, and whales hear according to the same principle as humans with the difference that the adjustment of acoustic air resistance to water resistance is eliminated.

The hunting techniques of bats

To the human ear, sound at frequencies above 20,000 hertz is inaudible. We can consider ourselves lucky that our ear is insensitive to ultrasound. Otherwise we would hardly get any sleep in many rural areas, for ultrasonic impulses are the navigational signals by which many species of bats hunt insects in the nocturnal skies. The sonar energy they emit in the process is considerable. At a distance of 5–10 centimeters from the head of a bat, the acoustic pressure can be as high as that of a pneumatic hammer. That much sound is required for the bat to get sufficiently clear reception of echo signals.

There is a good reason why bats use the ultrasound range: In order for sound to be effectively reflected by an object, the size of the object must be close to the wavelength of the sound. Many of the insects bats hunt are 3–5 millimeters in size. In order for the wavelength to be equally small, the frequency must lie between 11,000 and 66,000 hertz.

From the way bats vary the frequency of their ultrasound volleys while approaching an object, we have learned that wires 1 millimeter thick can be located from a distance of 2 meters. Wires 0.2 mm thick can be sensed at 90 cm, and 0.1-mm wires can just about be out-maneuvered. To thinner obstacles bats are ultrasonically blind. Infrared movies of hunting bats show the drama and intensity of ultrasound hunting. The insects flee in wild zigzag paths and use, in the smallest space, any conceivable trick to shake off their attackers. The bats often turn somersaults and get into bizarre flying positions while approaching an insect. When they come to within a wing's reach, they usually do not risk snapping for the insect with the mouth and missing it. Rather, they net it with the skin flap stretched between their legs or over their wing arms.

More details regarding the physics of navigation and range finding have become known from well-programmed experiments with captured bats. Not all bat families use the same technique. The largest family (*Vespertilionidae*), to which the much-studied mouse-eared bat *Myotis* belongs, hunts with very short ultrasound impulses. A *Myotis* looking for prey emits every 1/7,000 of a second an ultrasound impulse that lasts between 1/2,000 and 1/3,000 of a second. When it approaches the prey, the interval between impulses is cut down to 1/5,000 of a second and the duration of the impulse is lowered to 0.0003 second.

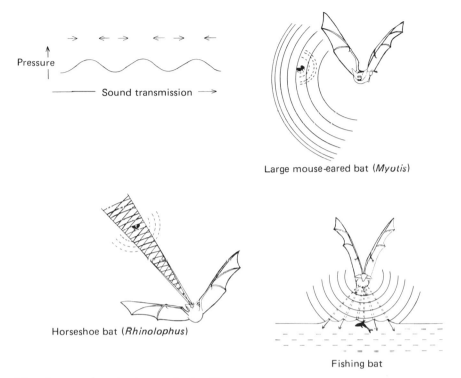

The ultrasound navigation methods of bats.

The reason for the shortening of the impulses is not hard to guess: An impulse lasting 0.003 second produces a sound wave of 1 meter. If the insect were less than half a meter away, the bat would still be emitting sound while the echo had already reached its ear, and it would not register a feeble echo with all that noise. The bat therefore reduces the length of the sound waves as it approaches its prey and operates in the end with wave bundles of only 10 centimeters. At the same time it increases the frequency of the emitted impulses, which enables it to register more quickly the acrobatic escape maneuvers of the insect. Modern radar systems that work with electromagnetic impulses operate in the same manner.

The horseshoe bats (*Rhinophidae*) use another method of ultrasound navigation. They emit the ultrasound not from the throat, but from the two nasal openings, a few millimeters apart, around which the horseshoe is located. The horseshoe is a structure that focuses the ultrasound like a concave mirror and directs it forward in a narrow beam. With this "searchlight," horseshoe bats scan their surroundings. They can focus the ultrasonic cone at will by muscular power. What is remarkable is that the emitted ultrasound impulses may reach lengths of more than a tenth of a second. The ultrasonic wave trains thus attain lengths of over 30 meters, which means that the echo must arrive while the sound is still being emitted. Apparently, horseshoe bats extract their hunting information from the sound pattern created by the superimposition of the emitted wave and the returning echo. The conditions are very favorable for such a process. Since identical sound is emitted from two openings about half a wavelength apart, the result in certain areas is total erasure of the sound waves by superimposition. It is conceivable that the sound-sensing devices are located inside such a quiet zone and can record the echo with relatively little interference. We have no precise details regarding the navigational mechanism of this type of bat.

Among the approximately 800 families of bats, there seem to exist innumerable intergrades between the two ultrasound navigational systems mentioned above. As varied as the frequencies and time patterns of ultrasound impulses are the tactical tasks of the echo direction-finding system. *Molossidae* always fly at a certain altitude, from which they execute spectacular nosedives after their prey. *Plecotus* and *Nycteridae* hover in front of foliage or branches and pick insects out of their resting places. *Megadermidae* are skillful hunters of mice, which they catch in daring chases over the ground.

Ultrasonic navigation is not an auxiliary device for night-blind hunters, but a mature acoustic sense.

In tropical and subtropical zones there are bats that live on fish they pull out of fresh water or the sea (*Noctilio, Pisonyx,* one species of *Myotis* from Asia). They fly in complete darkness very low above the water and emit ultrasound signals like other bats. From time to time they flop onto the water and pull a fish out with their hind legs. Their hunting technique is puzzling. Many a naval strategist has dreamt of being able to spot submarines by echo direction finding from the air. It doesn't work, and the reasons are quite clear to a physicist: Since sound travels faster in water than in air, it is refracted in a different way on the surface of the water than is light. If it deviates from the vertical angle of incidence by just 13°, it is completely reflected into the air. But with sound, 99.9 percent is reflected even when it hits the water's surface vertically. This has to do with the fact that acoustic resistance—the product of sound velocity and density—is very different for water and for air. Water acts like a very poorly adjusted microphone and absorbs little sound energy. Moreover, the acoustic resistance of a fish's body differs little from that of water. Hence, sound is probably reflected mainly by the air bladder of the fish. But because of the unfavorable acoustic adjustment, again only 0.1 percent can penetrate the water's surface in the direction of the bat. That is, the echo from a fish would be a million times weaker than the echo from a comparable object above water. It is inconceivable that a bat could hunt efficiently under such conditions.

The enigma started to unravel with more thorough studies of fishing bats' behavior. It turned out that they react in the first place to the ripples on the water surface, and grip any solid body that penetrates above the surface. This also explains why fishing bats like to hunt in the company of pelicans or schools of predatory fish, which cause many small fish to crowd in panic against the surface.

Underwater acoustic techniques

Echo navigation
Sounds of the most varied frequencies are, on the average, muffled almost 50 times less when passing through water than when passing through air. This means that acoustic signals can be transmitted in the ocean over considerably larger distances than in the air. Because sunlight is available only in the uppermost layers of the water, marine animals need another means for exploring their surroundings and communicating with other members of their species. Nature has been correspondingly generous with the application of acoustics in the oceans.

Whale hunters have long known that the seas are not silent, and science has delivered proof that the depths often reverberate with noise. The human ear cannot readily hear underwater sound, since it is adapted to a very different sound resistance in the air and absorbs only very little sound energy from the water. We usually need specially constructed underwater microphones. However, under favorable circumstances it is possible to perceive sound emissions of marine animals from inside a boat. I had such an experience on a fiberglass sailboat in the region of the Galapagos Islands. When a flock of dolphins turned up and entered a playful race with the boat, a merry chattering of chirps and whistles was audible inside the hull.

Dolphins produce whistling sounds and series of short clicks at sound frequencies of up to 170,000 hertz. They are unable to compete with bats in locating small objects, since sound waves are larger in the water because of the higher velocity there of sound. This probably is the reason why dolphins use higher frequencies. Still, the echo direction-finding system of the dolphins seems to surpass in quality the navigational system of the bats. In murky waters and with their eyes covered they effortlessly find their prey, and even blind they can distinguish dead fish from water-filled gelatine dummies.

In the pitch-dark antarctic or arctic winter, seals often dive for half an hour or longer from their ice-free airholes into the depths, and they obviously have no trouble finding their way back to the tiny opening in the ice. Penguins too manage surprisingly well in the wet darkness, though so far they have not been found to possess ultrasound senders. But there is hardly another possible way for them to navigate except by means of sound, which they may produce either deliberately or through their swimming motions.

How do sperm whales find their bearings when they hunt giant squid at a depth of 1,000 meters? How do they avoid ramming the ocean floor with the full force of their bodies when they dive? Evidently, numerous marine creatures recognize the contours of their surroundings with their eyes closed. The sound they produce serves them as illumination, as if they were carrying lanterns. Is it really so difficult to imagine having the ability to navigate with the help of acoustics? Actually, it is not. When we pass by a fence or through a tunnel by train or by car, we notice distinct changes in the noise level which are typical for the environment. Therefore, periodical fin strokes may easily act as secret senders for the sensitive sonar equipment of many aquatic animals. The sea harbors other acoustic secrets: Since sound is muffled only very little in seawater and the bridgable distance increases with the square of the acoustic wavelength, whales can in theory chat effortlessly in their deep tones over many hundreds of kilometers. Given their sparse numbers and their far and lonely itineraries, such communication would certainly be a great help for social contacts.

Sound amplification through vibrating gas bubbles
When you swim underwater and let air bubbles rise, you hear sounds. Such sounds are often heard in television shows about underwater exploration. Air bubbles produce sound because they are set in vibration when they are blown. Every time the surrounding water compresses the air in the bubble, the interior pressure increases and the water must again yield. In shallow water an air bubble 1 centimeter in diameter vibrates about 330 times per second. An air bubble with a diameter of 1 millimeter vibrates ten times as fast. The frequencies of these vibrations are distinctly audible to our ears, and just as distinct to the ears of fishes.

Vibratile bodies are set into especially strong vibration when they are periodically stimulated in the zone of their natural oscillation frequency. This phenomenon is well known in connection with radio. Good reception from a weak sender is obtained only when the receiving circuit is adjusted to the frequency of the sender. An automobile can be set into dangerous vibration when it is driven over undulating ground. An engine may actually be destroyed when it goes into covibration.

When gas bubbles are hit by a sound wave of a frequency close to their own, they start vibrating most intensively, while optimally extracting energy from the sound. Ths is just what happens in the air bladders of fishes. In the proper frequency ranges they can amplify sound waves a hundredfold. However, in order to make use of this amplification effect, the fish needs a suitable acoustic link between air bladder and ear. Many fishes (*Clupeiformes*, among them the herring) have anatomical processes which reach from the air bladder to the ear. In the ears of the fish family *Mormyroides* there are gas bubbles which have become completely detached from the air bladder. The majority of freshwater fishes have a chain of small ossicles to conduct the amplified sound from the air bladder to the ear.

The little trick of the vibrating gas bubble is well worthwhile. If the pressure fluctuations of the sound wave are increased a hundredfold, the fish can hear ten thousand times better. Now, since air bladders do act as resonators, why not use them to produce sound? This is indeed done by several fish species, for instance the sea robin, the freshwater drum, and the bass. They have developed drumming muscles which grip the outside of the air bladder and make it vibrate. Sometimes the air bladder has strands of tissue on the inside which can regulate the tension of the air bladder and thereby its transmitter frequency.

The acoustic navigation of the oilbird

In dark rock caves of the northern and northwestern regions of South America, there dwells a large hawklike bird, the oilbird or guacharo (*Steatornis caripensis*). It leaves its cave only under cover of darkness to feed on jungle fruit. The oilbird was described in Alexander von Humboldt's records of his famous journey through South America. Ever since, it has aroused the interest of zoologists because it has no close relatives and had to be classified as a family unto itself. In the mid-1950s this interest intensified when it was discovered that the bird was able to find its way in the darkness by means of audible acoustic signals.

The mysterious nocturnal way of life of oilbirds aroused my imagination. At last I found a cave inhabited by them on the jungle-covered eastern slope of the Peruvian Andes near Tingo Maria. When I entered, leaving behind me the oppressively humid heat of the jungle, I was welcomed by a pleasant coolness and silence. I sat down and rested on the soft dry floor, which was covered with nuts, seeds, and dried fruit peels with a few feathers mixed in. Slowly my eyes adjusted to the darkness.

In the front part of the tunnel, which was about 15 meters high and 40 meters wide, I saw no sign of the oilbird except for those remains of food and the feathers. A flock of small parrots played about the shady entrance to the cave. Some 200 meters farther in, after passing around a bend and finding myself surrounded by complete darkness, I was startled by the cries and the wing noise of large birds. Still farther inside the cave the noise became bloodcurdling, and I began to understand why the Indians who accompanied von Humboldt had refused to enter the cave of the oilbird. The composition of the sounds emitted by the birds sounded very complex. A raucous cry of alarm that can best be transcibed as "graw-graw" was particularly obtrusive. In between, there were sharp and frequent clicks uttered at irregular intervals, often several times a second.

In the light of my torch, or when I used my photographic flash, I saw flocks of large oilbirds fluttering excitedly about the cave. They usually kept close to the ceiling, but sometimes they came so close that I could feel the air current of their wings. It was evident that the birds were flying with complete certainty in the pitch-black cave. Every time I let my flash explode in the darkness after a lengthy pause, I could see a few birds flying around. There never was the slightest indication of a bird encountering any difficulties in navigation. My flash photos show that the oilbirds always maintain a proper distance from each other during flight. In total darkness this is quite a feat, and it can only be explained by sonar navigation. (To be sure, oilbirds fly rather slowly and cautiously when they navigate by clicking sounds, as can be seen from their flight patterns in the accompanying flash picture.)

The oilbird has a wingspan of up to a meter, and the wings are extraordinarily wide. The wing hands and tail feathers are widely fanned out, and the forewings are extended. The eyes are very strongly reflective—an indication that they are admirably adapted to seeing at night. In fact, oilbirds can comfortably find their way with their eyes and without the help of clicking sounds when there is sufficient light. But in complete darkness they must rely entirely on the short

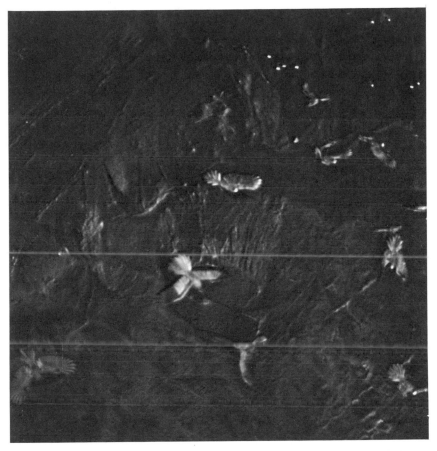

This flash picture, taken in total darkness, shows oilbirds in echo-guided flight. The strong light reflection of their eyes proves that they are equally well equipped for seeing in twilight.

sound impulses they emit and the echoes they receive. We know today that these sounds have an average duration of about 0.001 second and a frequency of about 7,000 hertz. Just as with some species of bats which navigate by ultrasound (for instance, *Myotis*), the sound impulses of the oilbird are so brief that the wave train of the traveling sound is less than a meter long. This means that during the subsequent period of silence the bird is able to perceive echoes from objects that are less than half a meter away. The wavelength of the emitted sound—almost 5 centimeters—is such that the oilbird cannot locate small objects in the range of a few millimeters; however, its direction-finding system is entirely appropriate to its leisurely life as vegetarian and cave dweller.

Muffling of sound and acoustic defense

The wings of a bird, with their complex air currents, their turbulences, and their vortices, generate sound waves audible to the human ear. For owls, which hunt mostly at night by acoustic direction finding, the noise of their own wings in flight constitutes the limits of perception. However, by some remarkable technical tricks which we do not quite understand yet, nature reduced to an astonishing degree the noise owls make in flight. In the plumage of an owl, all potentially turbulence-producing edges and projections have been carefully eliminated. Fine down on the top side of the feathers prevents noise when the feathers rub against each other. The frontal edge of the first feather in the wing has a fine comblike structure, which presumably has a favorable effect on the boundary layer of the wing surface. Very long, fine hairfeathers which are not interconnected and which flutter behind in the air current also have a muffling effect.

For moths, too, noiselessness is a weapon for survival. Though they are protected against being sighted by their nocturnal way of life, they are exposed to the sophisticated acoustic detection methods of specialized predators. Many moths have fine fringes at the tips of their wings which prevent the formation of the air turbulence that is responsible for flight noise. Just as cleverly designed are the wings and bodies of nocturnal lepidoptera, with their plushlike surfaces. The air particles in the fine pores of the soft layer are set in covibration when struck by a predator's sonar waves. By friction against the many obstacles, sound energy is quickly converted into heat and can no longer be reflected. The following acoustical data demonstrate how efficiently these lepidoptera can cut down ultrasound and render it harmless: Five-millimeter-thick hairfelt reflects sound at 500 hertz with

Moths' thick coats of hair can very effectively absorb and eliminate ultrasound from bats. This splendid specimen was photographed in Bolivia.

18 times less efficiency than does a concrete wall. At 2,050 hertz this reflection is 26 times poorer. Moths are able to listen in on the ultrasound impulses of their enemies. The ultrasound generated by the click of a wine glass at a garden party may cause a swarm of moths to drop to the ground like lead pellets.

5

Survival in Extreme Heat and Cold

Our sense of temperature tells us how different it feels to be hot or cold. Except for this subjective feeling, however, there is no physical contrast between hot and cold; it is only a matter of gradation. Heat, as we well know, is a form of energy: the kinetic energy of atoms and molecules. Temperature is the measure of the average heat energy within a system. We owe the heat or kinetic energy of our environment partly to the heat reservoir inside the earth, but mainly to irradiation from the sun. By interacting with matter, the energy of visible light and of infrared radiation is converted into kinetic energy of atoms and molecules, which can, in turn, produce and emit light and infrared radiation if their kinetic energy is high enough.

Two bodies of different temperatures thus exchange radiation until their temperatures are equalized. Black objects, which optimally absorb energy radiation of any kind, are also able to radiate it most efficiently. But if matter—for instance, air or water—is placed between objects, their temperatures are additionally equalized by kinetic energy traveling over their atoms and molecules. The rate of such heat diffusion depends greatly on the composition of the heat-conducting medium. Metals are very good conductors because electrons can move freely in them. In insulating materials, such as wood, the molecules are poor transmitters of kinetic energy because they lack mobile heat carriers. Tiny cavities in wood form additional obstacles to heat transmission. Wood therefore has a low coefficient of thermal conductivity. The transfer of thermal energy between two bodies works better the larger their difference in temperature, the smaller their distance, and the larger the thermal conductivity coefficient of the medium between them. Air is a much poorer conductor than water, since it has a much lower density of heat-carrying molecules. In gases and fluids, additional heat conduction occurs by currents or convection produced by temperature-dependent pressure differences.

Life as we know it is possible only within a very narrow temperature range. The kinetic energy of molecules in an organism must be sufficiently high to allow vital biochemical reactions to take place. On the other hand, an upper limit of molecular energy must not be exceeded, lest the complex proteins and the fine tissue of biological structures suffer damage.

The majority of mammals regulate their body temperature between 35 and 40°C. Birds too need to maintain high body temperatures (38–43°C) because of the high energy requirements of flight. By contrast, the body temperatures of most marine animals (other than mammals), reptiles, and insects fluctuate greatly according to environmental conditions. Mammals and birds are able to maintain activity in great heat or cold because they can regulate their temperature, but reptiles exhibit their full vitality only in the higher range of outside temperatures. When the temperature falls, they become lethargic and then rigid.

In the arctic and antarctic habitats of many animals the outside temperature may go down to −50°C or lower. In desert regions inhabited by animals the thermometer may rise to above 60°C. In the Kalahari Desert of South Africa the temperature can reach 70°C during the day and fall to −5°C that night. Animals can cope with such extreme environmental conditions only by sophisticated mastery of thermophysics and cryophysics.

The struggle against cold

The tale of an extraordinary drama of nature has come from Patagonia, at the southern tip of South America. In the year 1903, a particularly cruel winter dropped temperatures below −30°C. Thousands of guanacos (wild miniature camels) gathered in an area of a few square kilometers in the loop of a river. They crowded together and perished; those that were still alive climbed on the corpses, until in the end the dead bodies formed a huge hill. The surviving animals were so tame that one could walk among them as through a herd of sheep and had to push them aside to get ahead.

Patagonia is swept by strong winds and storms. The heat loss an animal suffers is considerably higher when the wind blows than when there is no wind because more heat is conducted away from the body surface. At a temperature of −30°C and a wind velocity of about 35 kilometers per hour, the heat loss corresponds to an environmental temperature of about −50°C. Under such conditions unprotected flesh freezes within a minute. Evidently, that was the temperature at which the slender guanacos, not well protected against the cold, met their icy death. In their distress, they probably tried to find relief in each other's body heat.

In the Yukon territory, at the edge of arctic Canada, temperatures of −50°C are no rarity. Large animals as well as many small ones have learned to survive this murderous cold without damage.

Large animals, like the caribou and the elk, have, from the viewpoint of physics, a better chance of coping with extreme cold. The size of their heat-emitting surface in proportion to their heat-producing volume is within tolerable limits. It is well known that animals of the same genus grow larger in size the farther they live from the equator and the closer to the polar regions. As an impressive example, penguins grow to only 50 centimeters on the equatorial Galapagos Islands, while the emperor penguin of the shores of Antarctica reaches a size of 1.20 meters. The largest hummingbird, the Patagonian colibri (up to 20 centimeters long), ventures south as far as the glaciers of Tierra del Fuego.

Smaller creatures, such as foxes, wolves, and rodents, compensate for the disadvantages of their large surface/volume ratio by burrowing into the lower snow layers. Since snow is a poor heat conductor, they usually find there temperatures only a little below freezing. The most important means nature has provided to animals for minimizing heat loss are well-insulating protective covers of hair or feathers, which usually are further improved by underlying layers of fat. The thickness of the fur reaches 6.5 centimeters for the Arctic wolf, and almost as much for the reindeer and the caribou. The excellent heat-insulating effect of fur and feathers derives from the many tiny air pockets they contain. As air is a very poor heat conductor, there is hardly a chance for heat to be transported by air currents inside fur. The thicker and denser the fur, the better the insulation it offers.

One might suppose that, in order to diminish heat loss, nature would have shaped small arctic and antarctic animals into feathery or furry balls. Interestingly, this is not the case. For example, the small lemmings, which because of their unfavorably large surface ought to have a great deal of heat insulation, have a pelt only one-third the thickness of a wolf's fur. The fur of the snow hare is only half as thick. The thin legs of birds that inhabit cold regions seem mostly not insulated at all. Likewise, the slender legs of reindeer or caribou are covered with only sparse fur. It is not difficult to discover the physical reasons for the poorer insulation of small animals and thin body segments: It is impractical to insulate narrow pipes with thick insulating layers, or to wear very thick gloves with fingers. A thick insulating layer enlarges

These gulls on the shore of the Pacific in Chile have turned the narrow side of their bodies toward the wind. The temperature of their unprotected legs is maintained far below that of the body so as to keep heat loss down.

the surface of a small body to such an extent that heat emission only increases.

Nature has found other ways of coping with the problem: Heat emission was reduced by a drastic lowering of the body temperature in the affected parts of the body. Cooler surfaces radiate, or course, correspondingly less heat. Gulls have a body temperature of about 38°C, but in the webs with which they paddle through the icy waters the temperature hovers between 0 and 5°C. In an environmental temperature of −30°C a reindeer needs to maintain a mean body temperature of 37°C in order to survive. The thinnest parts of its legs and hooves, however, have a temperature of only about 9°C. A sled dog working in temperatures of −30°C has on the upper surface of its paw a temperature of only 8°C; on the bottom side it is down to 0°C. Little heat energy is lost from such exposed spots, in spite of poor insulation.

Moreover, when exposed to cold, many species (including man) can reduce the heat transfer to the surface by constricting blood vessels under the skin. This enlarges the cold-insulating layer of the body surface, drives blood back into the body, and lowers the skin temperature. Arctic and antarctic creatures are built so as to maintain lower temperatures in parts of the body that would otherwise radiate much heat into the surroundings. In large mammals this undercooled layer under the body surface can be up to 10 centimeters thick.

Nature has mobilized additional mechanisms against the cold. When animals raise their hair or fluff up their feathers in the cold, they enlarge the insulating layer. (Goosepimples are a vestige of this process in man.) Creatures living in cold zones have a heightened metabolism and therefore generate more heat. Trembling increases muscular activity and thus generates heat. Body appendages such as ears, tails, and legs are short. Animals at rest often roll up to reduce further their heat-emitting surface. Sometimes they gather in flocks and protect each other; examples of such behavior are the monarch butterfly of North America and the ladybug, which hibernate in dense clusters. It is of interest that small Arctic animals like the lemming have an easier time surviving temperatures far below freezing than surviving a thaw; wet cold soaks through their fur and displaces the many heat-insulating air pockets. Precisely as a protection against this deadly moisture, penguins have developed a very dense furlike hide in which the air pockets are so narrow and the surface of the feathers so water-repellant that it cannot be soaked through.

Heat conservation is especially critical for mammals living in the water, such as whales, seals, and dolphins. Because of its greater density, water conducts heat twenty times better than does air. As far as their massive bodies are concerned, these animals can insulate them sufficiently with layers of fat. The problem lies with the heat insulation of their large, thin fins. It really makes one wonder that in the arctic and antarctic regions, where whales find rich nourishment, they do not freeze because of their fins.

Since the fins transmit the power by which whales swim, they cannot be simply cut off from the blood circulation. When the swimming fins of these mammals were examined in this respect (Scholander), a heat-conserving mechanism was discovered which is amazing in its ingeniousness and simplicity. Each of the larger arteries in the whale fin through which blood flows outward is surrounded by a ring-shaped bundle of tightly fitting venous canals, through which the blood flows back to the heart. Such an arrangement functions as a heat-exchange plant operating according to the principle of the countercurrent. The warm blood of the artery gives off heat through the thin arterial walls to the returning cold blood of the vein and cools off considerably in the process. Blood reaches the fins after having been cooled and fulfills its task without loss of body heat. The countercurrent principle finds wide application in technology. Offhand, one could call this another of the physiotechnical subtleties which nature too has discovered, but it is conceivable that it was this simple thermotechnical device that first made it possible for whales to conquer the icy but nutritious arctic and antarctic seas.

The most successful pioneers of the cold are not the storm-inured penguins, not the polar bears which continuously drift around the Arctic on ice floes, and not the seals which plunge from their air holes into the ice-covered darkness of the arctic and antarctic seas. This distinction goes rather to the inconspicuous lichen, a symbiosis of algae and fungi that often clings to the bare rocks of its habitat like splotches of colorful or gray paint. Lichens exist in icy places high in the mountains (up to 7,000 meters). On the Antarctic continent, some 350 species of lichens have conquered sparse rock and stone sites that are free of snow and ice during at least a short part of the year. Many of these locations are so steep that snow does not cling to them; some are exposed and are kept ice-free by storms; some are thawed by the summer sun. Lichens have managed to get to within 400 kilometers of the South Pole. Along the coast of Victoria Land on the outer edge of Antarctica, where much of the research into antarctic lichen took place, the annual mean temperature is $-15.5°C$. There is not a single frost-free day during the entire year. The lowest temperature is $-48°C$. The algae that together with fungi constitute the community of lichen obtain their energy through the process of photosynthesis, as do green plants. They absorb sunlight, split water to release oxygen, and build their organic molecules with the carbon dioxide in the air.

We know today that the transformation of luminous energy into chemical energy involves innumerable steps of chemical reactions, which occur partly in biological membranes and partly in the watery solution that surrounds them. In order for the interreacting molecules to make contact at the most favorable points and to couple briefly for the reaction, kinetic energy is needed—in other words, heat. Therefore, all chemical reactions slow down when the temperature falls, until they come to a complete stop. Besides, because of the greater volume of ice, freezing water bursts delicate cell tissues. But evidently, as far as lichens are concerned, the laws of temperature to which biological matter is subject have been circumvented. It has been established with certainty that photosynthetic energy transformation can take place in lichen at temperatures as low as $-10°C$ and, with lower yield, down to -18 and $-20°C$. Since these results seemed improbable, they were rechecked by the most painstaking methods.

For instance, the path of carbon, which at such low temperatures builds up into high-grade biological molecules, was traced with radioactive carbon atoms and confirmed (Lange). The light-driven miniature power plants of lichen accordingly still function long after the water inside has frozen. All in all, even the lowest temperatures do not seem to bother lichens. Lichens have been cooled to -40, -70, and $-196°C$ (nitrogen in air liquefies at this last temperature). The photosynthesis process in lichens instantly resumes when they are supplied with their habitual environmental temperatures. Lichens can still convert energy when they are to a large extent dehydrated. Their extraordinarily tough and maintenance-free photosynthetic power plants make it possible for them to survive under the brutal conditions of the icy wastes. When they again breathe light after the long arctic winter or after years under snow, and when the vise in which the cold holds life is somewhat loosened, lichens hasten to stockpile energy and prepare for the subsequent period of night and rigor.

There is no explanation for lichens' extraordinary lack of temperature sensitivity other than that they have developed, from organic material, "solid-state energy cells" for their energy conversion. All water molecules participating in the conversion of energy would have to be linked to larger molecules that had already moved into their proper position and together constituted a solid body. Movement of atoms and smaller molecules through such a solid body is possible, but restricted. Solid-state elements that transform energy are widely used in modern technology. We need only think of thermoelements for measuring temperature, or of solar cells (made of metals or semiconductors, respectively). In contrast with electrical cells and storage batteries, which also transform energy, they do not contain any fluids.

It seems, therefore, that it was not man who pioneered the conversion of energy by applying the laws of solid-state physics. Future generations who will continue the thrust into the cold of space should be able to copy from the lichens an ideal energy cell which builds up nutrients and frees oxygen under extreme conditions.

Coping with heat

Heat is just as threatening to life as cold. Thermal movement of molecules at high temperatures damages delicate biological structures, and vital enzymes and proteins alter their functions. It is more difficult to find protection against heat than it is against cold, as people who remember hot, sleepless nights will confirm. Still, a great many creatures have learned to live with heat. There are algae that endure

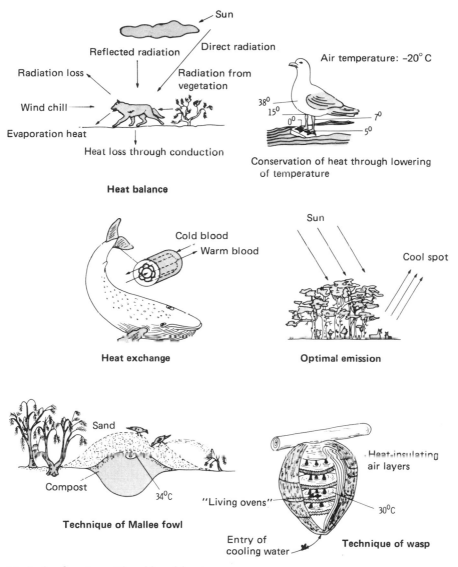

Sun

Reflected radiation

Direct radiation

Radiation loss

Radiation from vegetation

Wind chill

Evaporation heat

Heat loss through conduction

Heat balance

Air temperature: –20° C

38^0
15^0
0^0
7^0
5^0

Conservation of heat through lowering of temperature

Cold blood

Warm blood

Heat exchange

Sun

Cool spot

Optimal emission

Sand

Compost

34^0C

Technique of Mallee fowl

Heat-insulating air layers

"Living ovens"

30^0C

Entry of cooling water

Technique of wasp

Methods of coping with cold and heat.

temperatures of 85°C in hot wells and fishes that swim around merrily in 50°C water. Mosquito larvae (*Chironomidae*) develop normally in the hot springs of Yellowstone Park at water temperatures hardly tolerable to the touch. Plants too can endure great heat. When the sun burns down on desert plants, their surface temperatures may exceed 70–80°C.

Life in the desert is governed by the physics of heat. Whatever animals wish to survive adapt to it without reservations. The right size and shape can be of considerable advantage in a hot environment. Small animals with less body mass and more surface can stand heat much easier, since they generate little heat but can give off much. The long legs and big ears of many desert animals, among them gazelles and foxes, carry off heat efficiently. A suitable body position can also bring relief from the heat. Bats open their wings when they want to give off heat. Many animals (and humans) spread their limbs when they are uncomfortable in the heat. It seems, though, that the elephant with its big bulky body mass is an exception to the rule. In this case nature has tried another physical experiment: The size of the elephant is such that it needs a whole day to fully heat up. The big ears help keep the head cool in the process.

Animals that have learned to survive in the heat have little fat, and the body areas close to the surface are crisscrossed by a dense net of blood vessels through which heat is conducted to the body surface.

One of the most important physical methods of heat control is the evaporation of water. Water molecules need energy to detach them-selves from the semiorderly state of water and pass into the chaotic state of vapor. They extract this energy from the moist body. When a dog pants and puts out its wet tongue as far as possible, it achieves efficient cooling. (Carrying off heat by evaporation is of special im-portance to animals that live in cold climates and therefore wear dense insulating coats. When they get heated by overexertion, they could easily succumb to heat in spite of the cold environment.) Some animals even lick their fur when they suffer from heat. Perspiration on the body surface has precisely the purpose of making liquid available for evaporation. To allow sweat to evaporate, the humidity of the air must be low; this is why heat is less of an ordeal when the air is dry than when it is humid.

Many creatures seek water or moist places to get rid of heat. The tapirs of the oppressively hot Amazon lowlands appear regularly at waterholes to bathe whereas those that live in the high country of the eastern Andes do not have this habit.

The camel applies another principle of physics. It attains a favorable heat balance by raising its body temperature to above 40°C so that more heat can be given off to the environment by radiation and heat conduction. By raising its temperature, the camel uses the same principle used by animals that keep their body temperatures artificially low in the cold, only with reversed signals.

Animals that dwell in the hot desert often use several heat-control methods simultaneously, and also make very skillful use of whatever protection the terrain may afford.

Not every shade is equally cool. Hunters in hot regions usually know from experience that there is not much sense in pursuing game in the dense bush. During intense heat spells it is much more probable that the animals will be found on the shady edge of a glen or a trail. When an animal looks for a resting place, it is often not content with one protected only from the direct rays of the sun. Its fine sense of temperature leads it to those spots from which heat can be most efficiently given off. That is why animals often choose a shelter from the sun by looking for a place where they are shielded from direct rays and from which they can radiate their body heat without obstruction. When we look at the share long-wave infrared radiation has in the heat balance of an animal, it becomes clear how important it is for the animal to lose as much radiant heat as possible. Under normal temperature conditions a sheep, for instance, gives off about 48 percent of its heat in the form of radiation. It loses 38 percent by heat conduction, and only 14 percent through evaporation of moisture. Because of the large share of radiation energy in the thermal balance, animals protect themselves also at night against radiation losses to the cloudless sky—which may be as cold as $-80°C$—by resting under trees or rock ledges.

Reptiles in particular have developed a fine sense of temperature. They derive the major part of their body heat from the environment and have learned to optimally regulate heat reception and heat radiation by their behavior. Sometimes they make effective use of additional physical devices. Several species of lizards, for instance, change the color of their body surface according to whether they need to absorb more of less heat from the sun. The colors of many animals serve to either absorb or conserve energy. In tropical as well as polar zones the color white, or at least a very light color, is an advantage. In the tropics less radiation is absorbed by such coloring, and in the polar zones less heat is lost. As a matter of fact, in deserts and polar regions there are hardly any dark-colored animals.

Radiant heat is also one of the factors taken into consideration in many structures built by animals. Some ants and termites build narrow

Adaptation to climate accounts for the difference between the mountain lion of the cold southern Andes (left) and that of the hot Matto Grosso region (right). The former appears rounder, owing to its dense fur; the latter is thin-furred and lean, with darker coloring on the back to compensate for the shadow.

structures rising high above the ground which absorb a lot of heat radiation mornings and evenings and are to a large extent protected against it at noon. For the same reason, the meridian termite of Australia (*Amitermes meridionalis*) erects its structures (up to 5 meters high, 3 meters long, and very narrow) in a north-south direction. Here too evolution has given proper consideration to physical laws without their ever having been understood. The same applies to the intricate structural designs of cacti, which are exposed to a great deal of heat pressure in the desert. Their heat-reflecting capacity is low, since their surface is greatly reduced so as to cut down on evaporation. Nature has solved the problem by equipping many cacti with cooling ribs. These shade the cactus's surface against the scorching sun and simultaneously improve heat radiation. The alternating planes of light and shade of the vertical cooling ribs of the torch thistle produce rising and falling air currents, which improve heat radiation. And when the sun reaches its highest position, it hits the torch thistle from above, where it presents its smallest surface. A botanist discovered that torch thistles perish of burns when they are placed horizontally in the sun.

Active application of temperature physics in animals

The scrub fowl (*Megapodiidae*) of Australia, New Guinea, and the neighboring islands, of which there exist some 30 species, do not hatch their eggs with their own body heat. Rather, they build compost heaps of plant material and entrust their eggs to the heat of fermentation after insulating them with a cover of soil or sand. For this purpose the brush turkey (*Alectura lathami*) scrapes together a breeding mound 3–4 meters in diameter and 1.5 meters high above a prepared pit. The Australian scrub fowl (*Megapodius freycinet*) builds an incubator that is the largest of all bird structures; some are said to reach the incredible diameter of 12 meters and a height of 5 meters. By continuous digging, the bird can control the interior temperature of 33–35°C to within one degree, as the heat-generating decaying plant content is aerated or insulated. Scrub fowl evidently have an absolute sense of temperature. In order to take the temperature of the breeding mound, they dig a hole and test the compost with their beak. If the temperature is too high, they open ventilating shafts; if it is too low, they gather more fermentable material and improve the insulation of the heap by piling up more sand or soil. The mallee fowl (*Leipoa ocellata*), which inhabits the dry bush of Australia, works 10–11 months of the year without interruption to build its climate-controlled brood heap and to maintain the interior temperature for breeding in spite of fluctuating outside temperatures.

When an engineer wants to calculate or gauge the spatial spread of heat, he must solve the very complicated differential equation for heat conduction, which today most dare tackle only with an electronic calculator. Scrub fowl attack the daily supervision and reconstruction of their mounds as if the laws of heat distribution were entirely in their grasp. It seems as if, after establishing the interior temperature, they need only choose the proper profile of the breeding plant to maintain that temperature. The mallee fowl considers the existing climatic conditions instinctively (and, it seems, very sensibly) while it proceeds with its regulating activity. In spring, temporary air shafts are used to siphon off superfluous fermentation heat. When fermentation abates in summer and irradiation from the sun increases, the birds prevent overheating by piling up more sand and adding considerably to the height of the mound; they rely on inertia in the warming up of a large mass. Should nevertheless the heat of the sun penetrate dangerously deep, they change their tactics. They dig up the breeding mound early in the morning and spread the sand for cooling. When it has cooled off, it is again used to build the pile up. The spreading of sand as a method to further the reflection of heat into the air is a simple and physically very sensible procedure. Finally, when both fermentation and irradiation from the sun abate in the fall, the bird operates with a very thin layer of sand only, which quickly warms up in the sun. At the same time, sand is being heated in the sun close to the breeding mound under constant stirring; it is then mixed warm into the pile. When the Australian scientist Frith put heaters in the breeding heaps of mallee fowl, they reacted very sensibly. In the spring, assuming that the extra heat was additional fermentation heat, they ventilated the mound correspondingly more often. In the summer, on the other hand, they assumed that it was the sun that was heating their pile and built the heat-insulating sand layer correspondingly thicker, which caused the interior temperature to go up even more. In the autumn, when the outside temperature began to fall, the birds very sensibly omitted filling prewarmed sand into their mound when they considered its inner temperature to be adequate.

Some crocodiles and alligators have also proved to be surprisingly skillful heating engineers. The saltwater crocodile of south Asia (*Crocodylus porosus*) and the alligator of the southern United States (*Alligator mississippiensis*) build meter-high nests of moist branches, reeds, and leaves, into which they deposit their eggs. The female carefully watches over the nest, spraying it with water from time to time (by flapping her tail) to keep the fermentation incubator from going out. The Egyptian plover has a similar sense of temperature. It lays its eggs on the bare shady ground and covers them with sand to protect them against the sun. When the sand gets too hot, it flies to the closest body of water, fills its throat, and spits the refreshing liquid over the buried eggs.

Social insects too have developed remarkable climate-control mechanisms. In order to raise the temperature inside their nests, worker wasps are recruited to generate muscle heat. However, when it gets too hot inside the nest because of direct radiation from the sun, the wasps bring drops of water and wet the surfaces to produce cooling by evaporation. By this method they are able to cool their dwellings to below the temperature of the environment. The wasps have, of course, learned to keep the effort required for temperature control within limits by masterfully insulating their nests against heat. The building material they use, paper made of wood fiber and glue, is an almost ideal heat insulator in itself. In addition, they stack fine coats in such a way as to leave thin layers of air in between. Because they conduct poorly, these enclosed air layers retard heat loss. This is the same insulation principle that makes double-paned glass windows so effective.

Recently there has been much talk about the fascinating possibility of extracting heat from volcanoes or from the interior of the earth. Some gallinaceous birds have practiced this technique since time immemorial. Thousands of Australian scrub fowl gather at breeding time on an island in the western Pacific to dig breeding burrows into the warm, loose soil of a volcanic site.

Survival at High and Low Pressures

The dangers of diving

To man, diving is dangerous. In the course of our development of diving techniques, tragic accidents have dragged daring adventurers and entire submarine crews into a watery grave. The great depths of the oceans have been reached only with the help of complicated and not very flexible machinery which protects the diver against the immense pressures and provides the atmosphere needed for breathing.

Man cannot stay under water holding his breath for more than 2 minutes. Pearl divers usually dive for no longer than 90 seconds and reach a depth of 20 meters. In 1913 a Greek diver, S. Georghios, supposedly connected a rope to an anchor lost in 60 meters of water. Today's diving record of 70 meters for free diving should be difficult to improve upon to any appreciable extent.

With the help of breathing aids, these limits were considerably exceeded. It became possible to reach a depth of 180 meters by using a mixture of oxygen and helium for breathing, but to avoid critical injuries the diver needed more than 5 hours to equalize the pressure and carefully return to the surface. The problems encountered by air-breathing creatures under water arise not only from the lack of air, but also from the pressure exerted by water upon the air-filled body cavities and from the divergent physical properties of compressed air

The four main threats to which diving mammals expose themselves are concentration of carbon dioxide (CO_2) in the blood, inadequate oxygen supply, the water pressure on body cavities, and the problems that appear when the high pressure is lifted from the surfacing organism.

The lungs transmit oxygen to the blood and receive from it CO_2, which is eliminated. In a free-diving organism, CO_2 is gradually concentrated. If its share in the inhaled air approaches a value of 10 percent, man becomes apathetic and finally unconscious. Divers have two ways to cope with the CO_2 problem: by preventing the CO_2 level from rising too high, or by filtering the CO_2 from the pulmonary circulation. A safer method is the elimination of CO_2 by open circulation, which has been made possible by modern scuba equipment.

The faster a diving organism uses up the oxygen it has brought along, the shorter the acceptable diving time. The greatest danger is to the brain, because here an interruption of just a few minutes in the oxygen supply causes irreparable damage. Other parts of the body can be excluded from the oxygen supply for up to an hour without suffering any damage.

At great depth, an adequate supply of oxygen is not enough for survival; in order for the body to absorb oxygen, it must be inhaled at the same pressure which the water exerts on the body. However, the use of compressed air for diving has medical limits. When oxygen pressure inside the human body exceeds 1.7 atmospheres, dangerous cramps resembling epileptic seizures occur (the Paul-Bert effect). In modern diving equipment the pressure share of oxygen is kept artificially low (between 0.21 and 0.42 atmospheres) and the required total pressure is maintained by the less dangerous nitrogen or the noble gases argon or helium. But even these otherwise inert gases can become dangerous. According to Raoult's law, the solubility of a gas in a liquid increases with the partial pressure of the same gas in the adjoining gaseous space. When nitrogen is dissolved in the blood in too large a concentration, it causes the notorious "rapture of the deep," characterized by distinctive behavior disorders and occurring at depths between 60 and 80 meters. The inert gas argon has narcotic effects at 20–30 meters, whereas helium can be used at considerably lower depths.

Water pressure increases by one atmosphere or one kilogram per square centimeter with each 10 meters in depth. Since, according to the gas law, the product of pressure and volume remains constant, one liter of air is compressed to one-fourth its volume at a water depth of 40 meters. The same would happen to the air-filled cavities of the body—the thoracic cage and the nasal and tympanic cavities—insofar as this is anatomically possible. The dangerous pressures can be relieved only by an artificial raising of gas pressure in the body cavities, which can be done with compressed air or mixtures of oxygen and inert gases. However, breathing at great depth is considerably more laborious because of the increased viscosity of compressed gases, and in diving so much nitrogen is physically dissolved in the body tissues that it is released in the form of bubbles when a diver surfaces too quickly. The much-feared consequences are disorders in nerve functions and paralysis. Such injuries can be avoided by surfacing slowly and in several stages, and by patiently allowing the nitrogen the necessary time to escape from the tissues.

Diving mammals

When deep-sea cables had been laid across the oceans, unexpected and surprising information was gathered regarding great diving depths. In thirteen cases sperm whales have gotten entangled in these cables, which they may have taken for the tentacles of giant squid. Five such incidents have occurred at a depth of 900 meters. In one case, the cable lay at 988 meters under water. The physical conditions to which an air breathing creature is exposed down there are unimaginably extreme. The water pressure is about 100 atmospheres, which corresponds to a load of 100 kilograms on each square centimeter.

Somewhat more modest, if mighty compared to those of man, are the diving feats of other mammals. The Weddell seal manages to go down to 600 meters, and some of the other seal species reach 250 meters. (Most other diving mammals, birds, and reptiles confine them-selves to the top 20 meters.) What is amazing is the diving time some mammals are able to endure. The bottlehead whale remains submerged for up to 120 minutes, and the sperm whale has no need to renew its air supply for 90 minutes. It was possible to measure the diving time of the Weddell seal with great accuracy: Several specimens were taken to the solidly frozen McMurdo Sound on the coast of Antarctica and a hole was cut for them in the ice. There was no chance of escape. In its desperate search for a way out, one of the animals remained underwater for 43 minutes.

In the course of evolution, diving animals have undergone drastic specialization to withstand the dangers of diving. Large diving mam-mals, for instance, tolerate a high concentration of CO_2 in the body fluids without experiencing difficulties. The air they exhale after sur-facing may contain over 10 percent CO_2. Continuous oxygen supply to vital organs and the brain is safeguarded by cutting off from the oxygen supply muscle masses that are insensitive to oxygen. This is achieved by constriction of the blood vessels. At the same time, the pulse rate is lowered (in seals, as low as 10 percent of the normal rate). Diving animals convert, so to speak, into heart-lung-brain ma-chines. If one were to prick the skin or the muscles, no blood would seep from the wound.

This extreme automatic rationing of the oxygen supply in diving animals is the key to their performance and their insusceptibility to diving sickness. The air they bring along, which they use very sparingly, lasts for correspondingly long diving times. Since it is carried in limited amounts, there is no danger of too much nitrogen being concentrated in the tissues, as would be the case with an artificial supply of compressed air. Diving animals therefore do not run the risk of having nitrogen bubbles form in their nerves and brain when they surface quickly.

The problem of the compressive load of water, which is so serious for man, has found an elegant solution in nature. Bones, body fluids, and tissues are practically incompressible and are not affected by pressures of up to 150 atmospheres. But they transmit the pressure to the body cavities, the thorax, and the cranial cavities. As long as the pressure inside the cavities is equal to the outside pressure, the load is not perceived; but when no equilibrium is established there is danger of a collapse. However, no such misfortune can happen to deep-diving mammals. The front of their thorax is not enclosed by ribs, so their lungs can completely fold up under pressure. The small remaining cavities of the respiratory passages in the head are very stable structurally and are built so that under water pressure they can be automatically and hermetically sealed against the outside. The breathing hole of the bottlenose dolphin, for instance, can be plugged up.

The means by which animals have mastered the physics of great pressures are amazingly simple and elegant. But we only begin to properly appreciate these techniques when we consider that descending into the blackness of the deep sea is not a daring adventure for diving animals, but a necessary routine in their hunt for food.

Diving insects and spiders

Water-dwelling insects have developed methods of diving entirely different from those of mammals and reptiles. Since their body surface is large relative to the oxygen-consuming volume, the dissolved oxygen that travels through thin porous membranes from the water into their specialized respiratory system is in some cases sufficient for their requirements. Some insects (for instance, midge and stonefly larvae) have for this purpose developed large surfaces for the exchange of oxygen, but many must rely on the renewal of their oxygen supply above water level. Often they have to use clever tricks to get around the surface forces of the water, so as not to swallow it or be caught in the surface film. For example, the larva of the drone fly has a breathing tube (up to 12 centimeters long) that functions like a snorkel.

Large aquatic insects often bring along air from the surface, which can usually be noticed as a silvery film clinging to the underside of the body by a fine, water-repellent cushion of hair. Additional air supplies are carried under the wings or adhering as bubbles. The oxygen in these reserves would soon be used up if not for a special effect that comes to the aid of the insect: Since the entry of oxygen from the water into the gas bubbles occurs faster than the escape of nitrogen, the water-gas interface functions like an oxygen-porous physical membrane, which can considerably slow down the consumption of oxygen by the insect. Some aquatic insects have managed to stabilize these physical "gills"with the help of a cleverly designed and extremely fine coat of hair so that they need no longer emerge above the water surface to replenish their oxygen supplies. For this purpose the tips of the water-repellent fine hairs of the beetles *Haemonia* and *Elmis* are curved at a right angle so as to form a water-attracting surface which securely holds a thin coat of air underneath. The water beetle *Aphelocheirus* attains the same purpose with an extremely dense coat of hair (2 million hairs per square millimeter) from which the air cannot be expelled even at pressures of several atmospheres. Evidently, nature has not overlooked the fascinating physical possibilities offered by the small body size of diving insects.

A spider adapted to the water, *Argyroneta aquatica*, likewise carries its air supply like a silvery mantle around its dense coat of hair. However, this aquatic spider (which is found in Europe and Asia) seems to have tired of continuously having to come up for air; it has made additional improvements in its diving technique. By carrying air bubbles from the water surface down and gathering them in the shape of a balloon under a delicate web anchored to some water plant, the spider creates an underwater "base station" from which to set out on hunting forays.

Life at high altitudes

The ability of animals and humans to adapt to low oxygen pressure is surprising. At Chacaltaya the Bolivian Andes Club runs a ski resort at an altitude of some 5,500 meters. I was surprised to find that at this altitude skiing was still fun—at least as long as it was all downhill. However, when the lift broke down and the skis had to be hauled uphill, the sporting spirit soon evaporated. Tourists are well advised

A hummingbird engaged in aerobatics at an altitude of 4,000 meters above Lake Titicaca, Bolivia.

for the sake of their health to return before nightfall to La Paz, 1,700 meters lower.

The highest shepherd settlements in the Andes lie at about 5,400 meters. The highest work sites must be the sulfur mines situated at about 6,000 meters on the volcanos of the northern Chilean Andes. Only the *Indios* who are used to such altitudes can endure the extreme working conditions. Like other creatures adapted to heights, they have an exceptionally high concentration of oxygen-carrying red blood cells. Their respiratory muscles and cardiac activity are very strongly developed, and the total amount of pumped blood can be 20 percent above average. Much more energy is needed to pump such blood, which is more viscous, through the arteries. The expeditions to the summits of the Himalayas have demonstrated that well-trained men without breathing gear can climb to far above 7,000 meters without falling victim to acute mountain sickness or high-altitude collapse. British climbers have found spiders in the Himalayas at 6,700 meters. Members of an expedition to Mount Everest observed migrating geese (*Anser indicus*) at 8,800 meters.

7

Light Technology

Light in the visible range and in the somewhat long-wave infrared range has precisely the right energy to interact with the most weakly bound electrons in molecules. These electrons are the ones that determine the chemical bonds, and thus the chemical reactions, of molecules. Therefore, visible light can set off, or be produced by, chemical reactions. This was the energy prerequisite for the evolution of photosynthesis, of vision, and of biological light mechanisms.

In the course of evolution, light as a supplier and transmitter of energy has been put to manifold use for the benefit of living organisms. The physical properties of light had to be taken into careful consideration in the structures and functions of light-absorbing and light-producing biological systems.

Photosynthesis

If nature had not succeeded in exploiting the luminous energy of the sun for the synthesis of energy-rich molecules, there would be only very primitive life on earth and almost no free oxygen. The serious difficulties that stood in the way of effective conversion of light into chemical energy can best be assessed by those scientists who are still trying to decode individual steps of reaction in photosynthesis. The process of photosynthesis, which occurs in the power stations of green plants and some bacteria, is the only effective energy-converting mechanism of its kind known to us.

The capture of light quanta is an elementary process. All molecules are capable of absorbing light, if the energy of a light particle is just enough to push the molecule into a higher energetic state. For reasons of quantum mechanics, only very strictly defined discrete energy states are admissible, and the wavelength of the radiation must fit the light-absorbing molecule like a key fits the lock it is to open. The light-catching pigment of plants, chlorophyll, absorbs mainly red and blue light. The leaves of plants appear green because light of this wavelength is left over and dispersed into the surroundings.

The chemical energy of a molecule is determined by the distribution of its atoms and its electrons. The electrons with the weakest bonds are the ones that give the molecule its coherence by their negative charge and their orbits, which extend over many atoms. The same bonding electrons are also responsible for the absorption of luminous energy. With the energy they take over from light, they can frequent hitherto forbidden areas of the molecule, thereby adding considerably to its reactivity. When this happens, the light-absorbing molecule has passed into the excited state.

The big problem in the extraction of chemical energy from luminous energy is the instability of excited molecules. Just as a ball tossed high into the air must return to earth, an electron must return as quickly as possible to its energetically lowest and most stable state. In the process, the excess energy is released in form of light (fluorescence) or, with much more probablility, in the form of heat. Thus, whatever luminous energy were to remain would be nothing more than a some-what intensified molecular movement in the vicinity of the originally excited molecule. The result would be a loss of luminous energy. To perform a successful conversion of energy, it would be necessary to rescue the energy-rich chlorophyll electron within its extremely brief life span by transferring it into the energy-rich and stable electronic state of a suitable acceptor molecule. This is the critical primary step of photosynthesis, and it must be completed within a billionth of a second in order to uphold the effective level of energy conversion.

It would all be only half as complicated if the molecular partner's own reactivity were not enhanced by taking over the energy-rich electron. Within the briefest span it can again react with the electron-transmitting pigment, to which it thereby again loses the electron. It is therefore not possible to effectively convert luminous energy into chemical energy in a homogenous mix of absorbent pigment and electron-acceptor molecules. That is why nature has linked the process of photosynthesis to biological membranes with built-in electron acceptors, through which the excited electrons can efficiently be steered away from the highly reactive photochemical products and be separated from them.

The power stations of green plants are the choloroplasts, tiny cor-puscles about 1–10 micrometers (1 micrometer = 1/1,000 millimeter) in size. They are crisscrossed in the interior by much-pleated closed membrane systems called thylakoids. It is in these thylakoid membranes that the excitation energy of chlorophyll molecules is converted into chemical energy. The structure of this membrane is still under study. The light-absorbing chlorophyll molecules are probably embedded between an outer protein layer and an inner layer of fatty acid. Some 250–300 chlorophyll molecules form one photosynthetic unit and absorb luminous energy, which is transmitted to a single reactive chlorophyll molecule by nonradiative energy transfer. This molecule is able to deliver the energetically raised electron to a suitable electron acceptor, and to fetch an electron of lower energy from an electron donor and incorporate it into its emptied original state. One light quantum is not enough to separate an electron from a water molecule and make it available, along a long chain of electron-transmitting molecules, for synthesis of an energy-rich stable intermediate product (NADPH). That is why two light reactions are connected in series, and the excess energy is used for the purpose of producing the chemical energy carrier ATP, which together with NADPH and carbon dioxide is assembled in a homogenous reaction into carbohydrates (the Calvin cycle).

With an almost 100 percent quantum yield, the process of photosynthesis is optimally efficient. This could only be achieved by a sophisticated structural organization of matter. Only 2 percent of the available radiant energy is exploited, but that seems to be sufficient for the energy requirements.

The production of colors

Colors are fundamental in nature as signals of communication, lure, and warning, and as protective camouflage. Surprisingly, nature—which has done a great deal of experimenting with color—has three basically different methods for color production. These methods are applied with great imagination, side by side or in combination with one another.

The most widespread method is pigmentation. Animal pigments are mostly drops or grains containing the colored material which are embedded in the skin, the fur, or the feathers. When white light reaches them, part of the radiation is absorbed and the rest is reflected or let through. The mixture of precisely these remaining rays conveys the impression of color. Melanin pigments are in general responsible for the appearance of black and many brown shades in the animal world. Carotenoids, which are plentiful in carrots, produce red, orange, and yellow, but also the delicate pink of the flamingo. When bonded

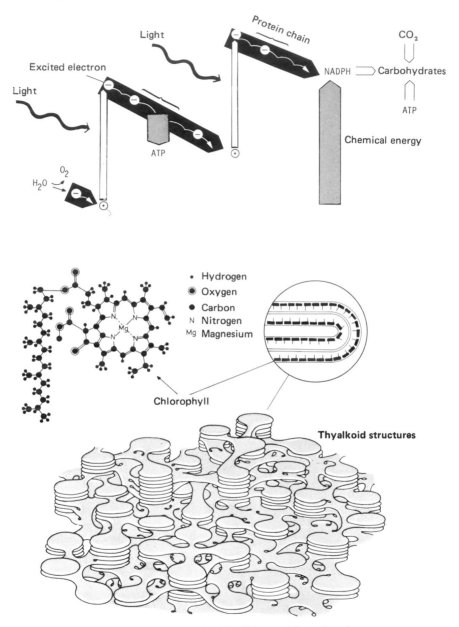

The process of photosynthesis. In principle, light particles raise electrons to higher energy states, at which they can initiate the buildup of energy-storing chemical products.

to proteins, they may produce blue and green. Lobsters owe their dark blue color to them: they turn the color of the unbonded carotene as soon as they get into the cooking pot, since carotene separates from protein when it is boiled. Many locusts and caterpillars get their rich green color from a combination of yellow and blue pigments. The olive-green plumages of many birds are due to a combination of black and yellow pigments.

As if chemistry were not flexible enough, nature has also developed two purely physical methods of color production. One of them resorts to light dispersion. It can be demonstrated that dispersion grows more intensive the more energy the light carries. Blue, energy-rich light disperses considerably better than red, energy-poor light. This explains why the cloudless sky appears blue against the black of space and why sunsets are red. Nature usually makes use of small protein particles as dispersion centers. If these particles are small enough, they disperse considerably more blue light than red light. Then, if we view the dispersed blue light against a black background, which may consist of the pigment melanin, we see a splendid blue coloration. This is the mechanism by which we get the color of blue eyes and the brilliant blue of the kingfisher and of parrots (called Tyndall blue after the scientist who first studied light dispersion in cloudy media). Blue eyes, therefore, can be called sky-blue in a strictly physical sense. Albinos have pink eyes because of the absence of melanin; the color of blood shines through from behind the iris. The brilliant green of many birds, snakes, lizards, and frogs is produced by a combination of Tyndall blue with a yellow pigment filter, through which the light must pass before and after dispersion. Whiteness in animals is likewise produced by the scatter effect. The size and distribution of the dispersing particles must be coordinated as in a cloud, so that light of all wavelengths can be dispersed. The dispersion centers are therefore larger, so that light of longer wavelengths can more easily take part in the dispersion. The white color of hair is produced by dispersion in tiny air pockets.

The other physical method leads to the creation of iridescent colors—interference colors which appear when white light is reflected from a thin layer or plate. They are produced by the different ways the waves are reflected in front and in back, which varies according to the different wavelengths of the light. Light that vibrates after reflection in exactly the opposite direction is eliminated. The remaining radiation produces the iridescence, which appears different depending on the visual angle to the reflecting surface. When the light strikes perpendicularly, those colors are extinguished for which the thickness of the reflecting layers constitutes an integer multiple of half their wavelength.

These iridescent colors can easily be observed in thin oil films on water puddles. Through precise engineering of the reflecting layers, nature manages to create intense iridescent colors. They are most exuberant in the plumage of the peacock. Pheasants and hummingbirds also adorn themselves with iridescent colors, as do many butterflies and beetles. These colors are produced on tiny platelets stored in animal cells in feathers, hair, or the chitin shells of insects. The silvery or multicolored scales of fish and the backs of their eyes are often extraordinarily reflective. In this case the reflecting structure consists of layers studded with guanine crystals. The spaces between the layers are filled with cytoplasmic fluid. The thickness of guanine crystals (0.001–0.002 millimeter) corresponds to one-fourth the wavelength of visible light they reflect. The secret of their high reflecting power lies in their thickness. The distance a light wave travels when reflected from a platelet one-fourth of a wavelength thick is exactly one wavelength, which gives optimal reflection. A shift of half a wavelength occurs because of the distance the light travels, and the other half of a wavelength is shifted through the reflection at the back of the platelet that borders on the optically thin cytoplasm. In other words, the physically optimal conditions for the reflection of light have found application here.

How light signals are picked up

The optical setup of a camera greatly resembles that of an eye; yet the makers of cameras have in no way copied nature. Rather, the similarity of these optical systems is dictated by the physical properties of light. It is therefore not surprising that lens-equipped eyes developed independently in essentially similar form in many families of animals— not only in vertebrates, but also in some jellyfish, bristle worms, snails, and octopi.

In bivalves, starfish, some snails, and many coelenterata, the visual cells of the retina are set in a cuplike pit. The light-sensitive visual cells, in which the visual impulses are stimulated by a photochemical reaction, are mostly separated by cells containing larger quantities of brownish-black melanin. In this way dispersed light is prevented from reaching them, and the direction of the incoming light can be fairly well recognized. The cup eye is the precursor of the bubble eye.

The evolution of the bubble eye brought considerable progress. A section of the surroundings is projected onto the retina as a reduced but sharp upside-down image. The transparent lens focuses incoming light through its specific geometric shape and its high refracting power.

Only because of this device can images and patterns be recognized and distinguished from one another by a scanning of the retinal image.

Vertebrates are able to bring the retinal image into sharp focus by the use of muscles. In order to project a close object onto the retina, either the curvature of the lens must be increased or the lens must be somewhat removed from the retina. The first of these adjusting methods is realized in humans and many other vertebrates, the second in amphibians. In osseous fish, the strong refraction of light in water has been taken into consideration. They have an almost spherical lens, which, though its refractive power is only slightly above that of water, still focuses a good deal of light. The eye of the fish is adjusted to close-range vision; only when distant objects are to be recognized is the lens shifted slightly toward the retina.

The difference in the eye's light-refractive power in air and in water has created difficult problems for diving animals. If man or one of the animals with air-adapted vision dives under water, the lens loses the refracting power and the eye becomes farsighted. This is why diving birds such as cormorants and ducks have developed a soft lens that can change its shape. It may turn almost pear-shaped under water to facilitate optimal vision. (The muscle tensions this requires are difficult to maintain over a long period of time.) Penguins, which spend longer periods in either element, are short-sighted on land and have normal sight in the water, where they find their prey and encounter their enemies. The small tropical fish *Anableps anableps* is specialized in catching insects running across the water's surface or flying close above it. The watchful eyes of this little fish that can be seen barely sticking out of the water are built for seeing in air. However, since the fish cannot afford to lose sight of its enemies underwater, it has a second pair of eyes below the surface, separated from the air eye by a dark horizontal band. Even nature was unable to come up with a more elegant solution for the light-refraction problems of this fish.

The structures of eyes vary with the requirements they must meet. Birds of prey and vultures have particularly sharp eyesight. To match the eyesight of a hawk, we would need field glasses with eightfold magnification. Songbirds too see much better than we do. In the Middle Ages, falconers often carried a shrike in a cage to guide them with its frightened looks along the path of the hunting falcon. Night predators such as owls, and creatures that dwell in the dim depths of the ocean, have eyes that collect a maximum of light in their large lenses. Usually they have another layer behind the retina which reflects light very well and so almost doubles the chance of a light particle hitting a visual cell.

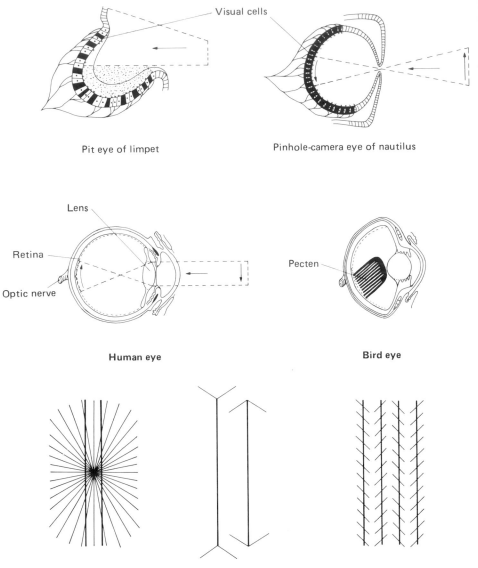

Pit eye of limpet

Pinhole-camera eye of nautilus

Human eye

Bird eye

Optical illusions with parallel lines

Nature has developed eyes into instruments of high optical performance. They register a few light quanta and resolve angular minutes, yet they may record gross optical illusions if the brain so wishes.

Most insects, and some octopi and bivalves (*Arca, Pectunculus*), have compound eyes, which in general are composed of a great many completely equipped photoreceptor units set next to one another, of which each captures only a small sector of light. Then the total image is put together by the brain like a mosaic to form an "autotype." This image is relatively sharp, since dispersed light is intercepted by pigment cells between the units. Of course, the sensitivity of the eye is reduced by this. Nocturnal insects can manage without this dimming device, and thus gain higher light sensitivity, but at the expense of visual acuity. When the light is stronger, many insects shift the screening pigment so that they get sharper vision. Dragonflies have screened and unscreened visual cells next to one another, and their vision is both sharp and sensitive.

The resolving power of a compound eye increases with the number of individual units. It probably reaches optimal strength in the large dragonfly, which has eyes composed of 10,000–30,000 elements. The scanning technique of the insect eye certainly carries with it a number of advantages. Insects can focus one visual unit, or a few of them, onto a target and fly toward it with hairbreadth precision. But they are also able to maintain a specific angle with respect to a point of reference (for instance, the sun). This type of scanning eye increases to an amazing degree the ability to see individually each one of many images following each other in rapid succession. Images that follow each other at intervals of 1/18–1/25 second appear to the human eye to merge, but good fliers among the insects can still distinguish images that follow each other at intervals of 1/100–1/250 second. This is important to small and lively fliers for control.

The perception of colors

The visual pigment of higher animals, rhodopsin, is a larger protein that contains a group of pigments and changes chemically over a cascade of intermediary steps when it catches the light. Some intermediate product stimulates a membrane and triggers a visual stimulus.

A few light particles are enough to produce visual perception, but to distinguish colors requires optical pigments that are sensitive to different wavelengths. The human eye can distinguish the colors of the rainbow from blue to red. Most of the solar energy that hits the earth is in this range of wavelengths (0.4–0.75 micrometers). Higher monkeys are sensitive to the same range of wavelengths. Birds, and many fishes and reptiles, are also capable of enjoying color. For a number of mammals, a certain measure of color perception is at least probable. Bees and other insects do not recognize red but are sensitive to ultraviolet light. In the sea of flowers which they navigate daily, the brilliant red of the rose or the poppy is missing, while other attractive shades show up that are hidden from the human eye because they are transmitted by ultraviolet light. There exist, however, red flowers that do not send out their long-wave color signals in vain; they attract hummingbirds and butterflies.

Many animals live in a black-and-white world, or recognize color only in a very limited range of wavelengths. Among them are numerous predators and insectivores. We must therefore be wary of looking for significance every time a warning, luring, or camouflaging color appears in a plant or an animal. It may be that the garish color of an insect is visible to the human eye only, not to the eye of the insect hunter. On the other hand, animals that appear well camouflaged in visible light may show very distinctive markings to the ultraviolet eye.

How light reflection is prevented in insect eyes

No optical system functions ideally. The different refractive index of air and cornea alone causes light to be reflected. When light falls vertically on glass, about 4 percent of the energy is lost by reflection. In more complicated optical systems, such losses would add up to considerable values. To reduce the loss, the surface of an artificial lens is lumenized by vaporizing a thin layer, which must be about as thick as one-fourth of the wavelength of the entering visible light and must have a refractive index between that of light and that of glass. This cuts light reflection considerably, since the reflecting waves for the most part extinguish themselves. However, when scientists inspected corneas of nocturnal insects through an electron microscope, they failed to find similar layers. Rather, the cornea was covered by a dense net of transparent cones, smaller than the wavelengths of visible light. For the purpose of studying the physical effects of these sugarloaflike attachments, an enlarged model was irradiated with microwaves corresponding to its scale (Bernhard). It turned out that the cones reduced the reflection of the radiation to a hundredth, and maintained this effect over a rather wide range of wavelengths.

The physical function of the conical attachments as coupling elements for the prevention of reflection is relatively easy to understand: In theory, a surface is optimally transparent if the refractive indices make a gradual and even transition from the value of one medium to that of the adjoining medium. The layer of cones on the cornea of insect eyes is constructed so that it has the refractive index of air where the

Nature has experimented with two different camouflage techniques in the South American guanacos: Countershading makes the shadow disappear, and the darker coloring of the head blends the animal into its surroundings

cones taper off, while at the base it reaches the refractive index of the cornea. All intermediate values appear in between. Because the cones on the cornea are smaller than the wavelengths of visible light, their geometric form is lost and the layer "appears" to light as an even zone with a continuously increasing refractive index.

This elegant device for preventing reflection gives nocturnal insects more intensive light perception. At the same time it prevents insects that during the day are camouflaged from betraying themselves by light reflections in their eyes.

The camouflaging of shadows

Nature has used colors with great imagination for the purpose of adapting animals to their natural surroundings. But of what use is the best camouflage coloring if, wherever an animal moves, it is accompanied by a dark shadow—possibly one that throws a sharp silhouette? Is not the attempt at ridding an animal of its shadow condemned to failure from the outset? Well, nature is patient and does not recoil from seemingly hopeless physical tasks. Experiments in this direction were remarkably successful.

One of the expedients resorted to consists in making an animal so flat that it melts into its own shadow. This method has been applied to the plaice and the ray. Both lie flat on the bottom of the sea in case of danger: the plaice on its side, the ray on its belly. In addition, they camouflage their contours with a bit of turned-up sand.

Many beetles and caterpillars have long lobular or hairy lateral appendices which they press against the ground to cover their shadow. For the same reason, moths lay their wings flat against the trunks on which they spend their days without moving. Butterflies sometimes lie on their side to cover their shadow. Often they align their folded wings with the sun, so that their shadow is only a thin line on the ground.

A body lit by the sun appears bright on top and dark on the bottom, where the shadow falls. Over a roundish animal, the transition from light to dark occurs gradually and allows the observer to get a clear impression of the body's shape. Nature counteracts this natural shading by painting the upper, brightly lit areas of many animals a darker color and saving the lighter colors for areas in the shadow, on the bottom. Through such countershading the animal appears as a shapeless, flat, and evenly colored object. This trick, which usually involves a gradation of the coloring, is common among deer, antelopes, camels, beasts of prey, and fishes. Animals camouflaged by spots and stripes (such as the big cats and zebras) usually have thicker and larger stripes or more numerous spots on top. This too is a way to produce a countershadow effect. Caterpillars may be darker either on the upper or on the lower side, depending on whether they crawl along the top or the bottom side of branches. *Synodontis*, a fish of the Nile that swims belly-up, has its darker shading on the belly, not on the back.

The ideal method of preventing the formation of shadows would consist in making animals transparent. Nature has experimented with this option too. An entire group of South American butterflies (*Pteronymia*) have crystal-clear wings. Not only do they not throw a shadow, but they are almost invisible. Cicadas too have transparent wings.

Antishadow technique reaches amazingly close to perfection in some fishes, among them the coral fish (*Coralliozetus cardonae*). Not only are they almost completely transparent; they even show distinct countershading on some of their inner organs, which for chemical reasons could not be made of translucent material.

Visual tricks

One of the most amazing achievements of evolution is the imaginative use of color patterns to protect animals against their enemies or cam-

This encounter in the Amazon region convinced me that the jaguar's spots imitate the fallen leaves on the jungle floor.

ouflage them from their prey. This task was by no means simple from the physical point of view, especially since it stipulated that the visible light reflected by the animal remain within a precisely limited range of wavelengths—that is, the surface had to be constructed so as to eliminate the rest of the light. Nature patiently threw the dice again and again in ever new color combinations until a certain pattern proved to be of particular advantage to a species and thus prevailed.

There are insects that look exactly like fallen leaves, marked on top with simulated leaf ribs. Sometimes they even look damaged or seem to have been gnawed at, like the leaf insect (*Phyllium*) of Malaysia. A number of mantises and butterflies also imitate leaves. Some of these butterflies even have transparent windows in their wings to make them look like the ribbing of a leaf. The South American moth *Draconia rusina* is quite safe from enemies thanks to this trick. Other butterflies imitate curled-up, withered leaves; this allows them a more relaxed posture.

Certain frogs and toads of the tropical jungle (*Bufo typhonius, Megophrys nasula*) also could be taken for leaves. The scalare, a fish of the Amazon region, not only looks like a fallen leaf, it also behaves like one when it drifts with the current. Numerous insects and several smaller vertebrates masquerade as twigs or pieces of wood; snakes sometimes as hanging lianas. Quite a few insects mimic lichen or flowers.

This Amazon bittern bears the vertical lines of reeds.

The clearly marked "front" of this Yucatan moth is actually the rear.

The lines across the body of this anteater separate the image of the animal into unconnected parts.

Camouflage by irregular spots is widely used. Such spots have nothing to do with the shape of the animal. Their purpose is to divert attention from the shape, or to let it merge with the surroundings. This kind of color camouflage separates the body of the animal into two or more unconnected parts. The patterns of snakes and frogs are examples of this method.

Many creatures wear the pattern of their surroundings. The bittern shows the straight thin stripes of the reeds in which it lives, and also assumes a corresponding posture. The tiger is adorned with the black and yellow coloring of tall dry grass, its favorite habitat, and the jaguar wears the yellow and black speckled coloration of the leaf-covered jungle floor.

Camouflage is only one of nature's visual games. Intimidation is another. It usually forms the second line of defense, when an animal has been discovered by a foe. Often this maneuver consists in the sudden showing of garishly colored imitation eyes. Eye spots are very common on insect wings. Some caterpillars, too, protect themselves in this way. I also noticed such eyes in back of the ears of a young ocelot. Obviously, they were to protect it against an enemy attacking from the rear. Other color signals are meant to divert the attacker from vital parts of the body. Some butterflies have mock eye and head sections at the tips of their wings. The southeast Asian beetle *Ancyra aunamensis* carries a conspicuous imitation head on its rear end, which it readily sacrifices to save its life.

Innumerable animals adorn themselves with the colors of poisonous or unpalatable related species. The rare and very venomous true coral snake of South America, with its conspicuous red, black, and yellowish cross-stripes, bears a close resemblance to a less poisonous but more widespread species. Other harmless snakes have the same pattern as a venomous species. Seventy-five species of coral snakes live on their bad reputation, but only few deserve it. In the jungles of South America, Africa, and Asia one can be almost certain to find for each poisonous or horrid-tasting insect many harmless lookalikes.

Sometimes the right camouflage also makes it easier to catch prey. The predatory fish *Aspidontus* mimics the cleanerfish, which is tolerated by larger fish and feeds on their parasites. Expecting a "cleaning," the victim allows the *Aspidontus* to approach. The next thing that happens is that the *Aspidontus* takes a big bite out of its unwary prey.

But for its impersonality, nature would have to be credited with a sense of humor, which also shows in the idea of making some insects look like bird droppings against the leaves of trees and shrubs. Some caterpillars look amazingly like semidried pellets of bird droppings. Even a few nocturnal lepidoptera imitate the shape and form of bird excrement. They rather resemble flattened blots, and owe their protection to the fact that they look like excrement dropped from great height. The crab spider of Borneo camouflages as bird excrement in order to prey on insects which sometimes look for nutrients in droppings.

Orientation by polarized light

When desert ants (*Myrmecocystus*) wander over the monotonous landscape of their hunting grounds, or when bees look for blossoms in the maze of shrubs and clearings, they need navigational aids. Thirty years after von Frisch's pioneering investigations into the life of the bee, we know that to many insects the sky does not appear as devoid of structure as it does to us. They recognize in it patterns which allow them to pinpoint their precise location without sighting the sun. With their intricate compound eyes, bees and ants are able to determine the plane of vibration of the electric field of a light source, and thus have access to a special kind of "sky compass."

The radiation that reaches our eye from the sky when the sun is covered is exclusively light that was dispersed in the atmosphere. During this dispersion process, the light waves of the sun, which vibrate in all directions with their electrical field, are fanned out by minute density fluctuations in the air and by tiny droplets of water and are sorted according to their plane of vibration. The vibration plane of the electrical field appears optimally aligned when the dispersed light is viewed perpendicular to the direction of incidence of the sunlight. In this case the electrical field vibrates at right angles to the plane formed by the incident solar radiation and the observer. The physical reason for this coordination of angles is easy to understand: The electrical field of incident light accelerates the electrons of molecules in its plane of vibration. But when the luminous energy is again reflected, it always does so preferably at a right angle to the accelerated electrons. On the basis of this knowledge, the to-us-invisible pattern of the polarized sky can be reconstructed.

We know that direct sunlight is not polarized, and that the polarization planes are set in circular form around the sun. A band extends through the sun and the highest point in the sky above the observer in which the dispersed light is polarized parallel to the ground. At a right angle to it is a band of light whose polarization plane meets the horizon at a right angle. However, one small visible spot in the sky does not suffice for complete navigation. In order to distinguish between the two identical halves of the sky, which are separated by the solar meridian, an insect needs further information. This information could easily be obtained by comparing the light intensity at two or more points in the sky, or by comparing the degree of polarization. Light undergoes stronger dispersion the shorter its wavelength. Ultraviolet light with about half the wavelength of red light appears sixteen times more intense in dispersed light. It therefore did not come as a surprise when it was proved that the polarization eyes of insects operate in the near-ultraviolet range.

How do insects perceive polarized light? The compound eye of an insect consists of a multitude of identical cone-shaped visual elements. Bees have 5,500 such units. Each is able to respond to only a small sector of the environment (a sight angle of about 5°), but together they scan a very wide field of vision. The individual vision elements, or ommatidia, function as independent light-transmitting photoreceptors. Each has its own lens and a crystalline cone which is lightproofed by

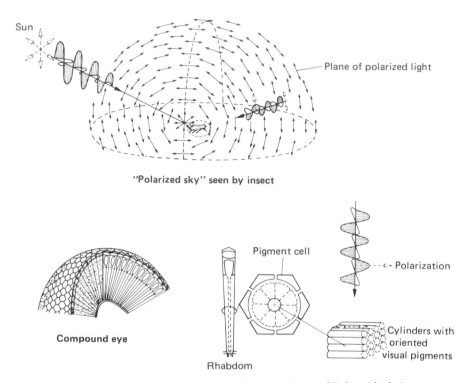

"Polarized sky" seen by insect

Compound eye

Rhabdom

Bees and many other insects perceive the vibration planes of light with their faceted eyes. To them, the sky is structured and gives directions even when the sun is covered.

a pigment screen. The cone transmits the light to a central translucent cylinder, which again is surrounded by eight or nine photoreceptor cells. The central cylinder possesses a number of optical refinements. Because its optical density is higher than that of the environment, it operates like a light conductor and does not lose any light to the surroundings. The principle applied here is similar to the one used in modern fiber optics. The central cylinder, with its eight subsections, contains in its lower region stacks of perfectly aligned tubelike structures which point with their openings in the direction of the corresponding photoreceptor cell. The optical pigment is stored in the walls of the tubes. Since the optical stimulus of a molecule occurs optimally in a specific molecular axis, a stimulus in the direction of the tubular axis would, because of the cylindrical geometry of the molecular pattern, be twice as probable as one at a right angle to it. There are additional indications that the optical pigment is aligned in the direction of the axes. The cylinders are subdivided into dichroic structures—that is, structures that absorb light from different planes of vibration to a different degree, just like polarizing glasses.

The Vikings used polarized light for orientation. On their far-flung journeys across the stormy and in winter very dark Northern seas between Vinland, Greenland, and Russia, they could not rely on sightings of the sun. Rather, they steered by polarized light, which they measured with the "sun stone," a disk with an oriented cordierite crystal (a magnesium-aluminum-silicate compound) imbedded in the center. A cordierite crystal has one light axis and lets polarized light pass through only parallel to the axis. When the sky was covered with heavy clouds, or after sunset, the Vikings held this stone up against the sky and turned it until clear light came through. After sunset the polarized light comes down vertically from the sky. From the plane of the light, the direction of the sun after it has set can be established at a right angle to that plane.

Bioluminescence

I first became aware of how fascinating the development and application of cold light in living organisms was when I camped one night in the jungle of the Paraguayan Chaco. I did not trust my eyes when I suddenly saw a miniature train with a red headlight and eleven pale green pairs of windows crawling along the ground in front of my face. I repeatedly rubbed my eyes, but it did not help. It was not until the glowworm crawled about in my palm that I stopped believing in a phantasma. I spent a good part of the rest of the night trying to shoot a few pictures of this colorful nocturnal visitor. This was by no means easy, since the glowworm was a zestful runner and also fiddled continuously with its lights. Soon it would have turned off the illumination altogether, but I discovered that it was possible to get it glowing again by blowing on it. At times it only turned off its red headlight, which shone as strongly as a burning cigarette. Later I found out that I had come across the female larva of the rare South American glowworm *Phrixothrix* (also called the "railroad worm"), which is unique among the many known glowing creatures in that it emits long-wave red light.

Biological glowing is a relatively common phenomenon. It occurs with most frequency among marine creatures. The host of lantern carriers extends from microorganisms (dinoflagellates) to worms, snails, bivalves, crabs, and fishes. The only known freshwater dweller that glows is the night snail of New Zealand. Among terrestrial creatures there are the luminous centipedes, worms, and insects, but no higher animals. Among plants, bioluminescence occurs only in bacteria and fungi.

Nature makes imaginative and varied use of biological light to lure members of the species or prey and to scare off foes. In the dark abysses of the seas, light patterns are a means of communication visible over long distances. Among the lantern fishes (*Myctophidae*) alone we find some 150 different species with different "neon signs." The black anglerfish (*Melanocetus*) proffers a luminous organ as bait on a rodlike appendix in front of its voracious mouth. One member of the family of deep-sea viperfish (*Chauliodus*) goes a step farther by illuminating the interior of its mouth. The flash is a common means of scaring off foes. How ingeniously light can be used as camouflage is demonstrated by the deep-sea hatchetfish (*Argyropelecus*). From its many luminous organs it radiates intensive bluish light downward and a softer light to the sides, so that it merges into the surroundings and becomes almost invisible in the diffuse light of the sunlit upper layers of water.

There are a great many known kinds of glowworms. In tropical regions they often flash characteristic sequences of signals at well-defined intervals so as to find their way among the swarms of glowing insects. Sometimes they produce grandiose spectacles when, like *Photius pallens* of Jamaica or *Colophotia* of Burma, they gather in a single tree in the hundreds or the thousands.

Incomparable in its effect is the marine phosphorescence created by the dinoflagellate *Noctiluca scintillans*, which I had occasion to observe in the waters around the Galapagos Islands. These creatures, about a millimeter long, emit a luminescence whenever they are stimulated by pressure fluctuations in the water. This is how it happens that luminous contrails are seen at night in the wakes of fish, that the crests of the waves and the surf fluoresce, and that the surface of the sea shines far and wide during a rainstorm.

In terms of energy, the conversion of chemical energy into luminous energy could be described as reverse photosynthesis. Today we know that the molecular conditions for these two processes are entirely different. It is much more difficult to transform in a controlled manner luminous energy into chemical energy than to produce cold light. Chemistry knows many light-generating chemical reactions that can conveniently take place in the test tube, but a controlled conversion of luminous energy into an energy-storing chemical compound is still far off.

The mechanism of bioluminescence in most if not all luminous organisms has been explained in rough outlines through the pioneering work of Harvey, McElroy, Seliger, and other scientists. The molecular "light carrier" is a small, rather complex molecule which occurs in various modifications in different creatures and which has been given the descriptive name of luciferin. When it is oxidized in the presence of the biological energy carrier adenosine triphosphate and of a complex and so far not analyzed enzyme (luciferase), it turns into an excited molecule which returns to its original state by emitting a light quantum. Glowworms lose during this reaction less than 1 percent of the energy as heat, which explains the expression "cold light." Other organisms attain less favorable yields. The emitted light is usually blue-green, yellow-green, or orange, but given the necessary conditions by the right electronic structure of the glowing molecule, it may also appear as the red glow of the "railroad worm." The luciferin reaction is not the only mechanism in nature that produces cold light. The blue glow of the medusa *Aequorea aequorea* is due to the light protein aequorin, which is brought to luminescence by calcium ions.

Infrared-sensitive snakes

When night falls over the Amazon region, the natives shun their familiar paths and the jungle becomes the realm of snakes hunting their prey with deadly accuracy in the deep darkness. Their guide is low-energy infrared radiation, invisible to our eye, which is emitted by warm living bodies and which they perceive with their highly sensitive receptor organs. Coloring, camouflage, or holding still is of no avail; the prey is distinguishable from its surroundings by its body temperature. To the snake hunting by infrared light it is a distinctly glowing spot.

Two of the fourteen snake families, the boas (*Boidae*) and the pit vipers (*Crotalidae*), have developed receptors for heat radiation. Both are major families. The former comprises the boas and anacondas of South America and the pythons of Asia. To the latter belong the very venomous rattlesnakes, the bushmasters, and the copperheads, all of the New World. While the human eye is sensitive only to light waves between 0.4 micrometer (violet) and about 0.75 micrometer (red), the receptors of snakes, in order to register heat radiation of mammals and birds, must react to radiation between 5 and 14 micrometers. Experiments have shown that the infrared organs of snakes can register low-energy radiation in this range of wavelengths with almost unbelievable sensitivity.

For pit vipers the limit of sensitivity lies at a radiant intensity of 0.0003 gram-calorie per square centimeter per second. This corresponds, per second, to the quantity of heat needed to raise the temperature of one cubic millimeter of water by 0.3 degree. However, a

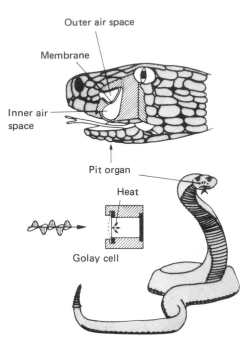

Outer air space

Membrane

Inner air
space

Pit organ

Heat

Golay cell

The infrared organs of the rattlesnake.

pit viper needs this radiation during less than a hundredth of a second
to recognize the signal. This means that snakes of this group are able
to distinguish in complete darkness without any trouble an object close
by that differs by only 0.1° from the temperature of its environment.
The infrared sensitivity of the boa is somewhat lower, but it is more
than adequate for efficient nocturnal hunting of warm-blooded animals.
The heat-sensitive receptors of pit vipers, which lie between the eyes
and the nasal openings in the pit organs, consist of two cavities separated
by a thin membrane. The outer cavity is connected with the outside
by an opening directed forward, in the snake's line of sight. In the
membrane are located the nerve endings that register the heat radiation
and pass the information on to the brain. In one kind of boa, pits
containing heat receptors are found on the lips.

Gamow and Harris did not find any indication that heat detection
occurs in snakes over a photochemical mechanism similar to the eye.
This would have been rather improbable, since molecules which are
decomposed by such long-wave infrared radiation would be unstable
at normal temperatures. Neither is it conceivable that a change in
temperature of only 0.003° in the membrane of a snake receptor,
which already can trigger a signal, would cause a well-defined chemical

reaction. A suitable thermal mechanism for the infrared organs of snakes has not, so far as I know, been suggested.

From the viewpoint of physics, only one principle presents itself that allows for sufficiently sensitive infrared recording without the need for cooling the infrared detector itself. This principle is put to work in the so-called Golay cell. The Golay cell consists of a thin membrane which absorbs heat radiation and with this absorbed energy heats gas in a small adjoining chamber. The flexible membrane is deformed as gas pressure develops and thus serves as a gauge for the absorbed energy, which can be measured electronically or optically. Superficially, the Golay cell looks amazingly like the heat-sensitive organ of the pit viper. It also has the same reaction time (0.002–0.03 second); an entirely adequate sensitivity given the environmental temperatures; and an extraordinarily broad area of sensitivity, reaching from ultraviolet radiation to microwaves. Such characteristics were also found in boas. Does nature resort to the same clever device and actually register heat-dependent pressure fluctuations? Someday nature may help us discover a completely new principle for measuring infrared radiation.

8

Electricity

Nerve transmission

Life began in the sea. Therefore, the physical and chemical properties of seawater laid out the path of development for much of life's physical equipment. This applies in particular to bioelectrical phenomena. Water molecules, because of the rather uneven charge distribution of their atoms, exert considerable forces. That is why many chemical elements are stable in water only in charged form—that is, as ions. Salts dissolved in seawater disintegrate into positive and negative ions, which are distributed so that their charges shield each other and the water in its entirety remains neutral. The presence of ions turns saline water into a remarkably good electrical conductor. If ions with positive and negative charges are spatially separated, it is possible to build up electric potentials and drive with them ionic currents.

Life has not come up with any structures capable of generating metallic conduction through migration of electrons. All bioelectrical phenomena known so far can be attributed to the electrochemistry of ions. In this respect nature has attained amazing perfection. The triumphal march of bioelectrical systems—the biological transducers—started with simple electric potentials which originated in the fact that biological membranes have different permeability for different ions. It led, over the development of the neural system and the complex feedback controls of the sensory system, to the construction of a computer unsurpassed in its flexibility: the brain. Moreover, nature has succeeded in developing an electrical direction-finding system which has not been paralleled in human technology. The most fundamental and most momentous bioelectrical achievement of nature was the development of structures capable of producing and transmitting electrical signals.

It was the membranes of the nerve cells that became carriers of electrical signals. They are equipped with a long tubular appendix—the nerve fiber, or axon—which can be electrically stimulated and serves as a cable for the transmission of electrical impulses. The decisive bioelectrical processes happen in its membrane, which is between 50 and 100 angstroms thick (one angstrom = one millionth of a millimeter). Thanks to very thorough research into the giant nerve fibers of octopi, which are up to one millimeter thick, we know in principle how a nerve impulse is triggered and transmitted. We are, however, still in the dark regarding many details of the processes occurring inside the thin nerve membrane itself.

When at rest, the nerve membrane is charged with 90 millivolts, so that the interior of the nerve fiber constitutes the negative pole. The nerve cell attains this charge by pumping positively charged sodium ions outward and a different quantity of potassium ions inward. If this rest potential is reduced to less than 45 millivolts by some external influence (for instance, some electrical, chemical, or mechanical disturbance), the electrochemical equilibrium at the nerve membrane collapses. Suddenly the membrane becomes permeable to sodium ions, which start penetrating into the interior of the nerve fiber like an avalanche. Accordingly, the membrane potential is further lowered, and this further increases the permeability of the membrane to sodium ions. Now the membrane begins to be permeable also to potassium ions, which flow outward. In the course of this ion exchange the membrane potential collapses altogether and even overshoots into the positive. Thus the maximum potential, the action potential of the nerve impulse, is reached. It amounts to about 0.1 volt. The ducts through which the sodium ions flow into the nerve fiber now begin to seal off, and the pendulum swings back to the original rest potential of the membrane. The entire process of excitation is over in a few thousandths of a second—but only in a narrowly limited spot. The electrochemical disturbance of equilibrium spreads to areas adjoining the nerve membrane and causes a collapse of potential there too. In this way the nerve impulse is transmitted. The nerve fibers transmit it at a speed of about 1 meter per second. The very thick fibers of octopi have much less resistance and transmit impulses at a speed of 30 meters per second. The only trace a passing electrical impulse leaves behind in a nerve fiber is the addition of a small quantity of sodium ions and the loss of potassium ions. A nerve fiber could transmit without any problem thousands of impulses before the change in the original concentration of ions would be enough to rob it of its ability to function.

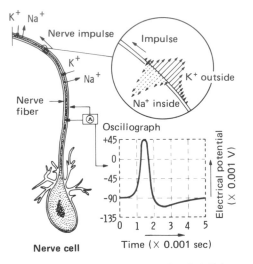

Nerve cell   Time (× 0.001 sec)

A nerve impulse originates with a brief change in the permeability of a nerve membrane; positive sodium and potassium ions pass through and temporarily shift the membrane potential.

Ion pumps built into the membrane see to it that sodium and potassium ions are continuously exchanged.

Measurement, control, and data processing

Optimal mastery of the physical laws notwithstanding, life would have been stymied in its growth at an early stage if not for the development of complicated systems of information, control, and data processing. One look at the human eye and how it functions makes this very clear: Some 100 million visual cells supply an abundance of information which would be unusable without classification and evaluation. Like the nerve signals of other sensory organs, they consist generally of sequences of electrical impulses. Their intervals vary greatly and are a measure for the physical and chemical changes triggered inside the sensory cell. Already in the eye the channels of information are reduced to 1 million by an ingenious process of joining and connecting in opposition. Then they are transmitted to the brain by the optic nerve. The nonstop salvos of impulses that bombard the brain must be processed according to a very intricate system so that quick reflex and command reactions can be triggered. The computer center in the head is so constructed that, on the basis of stored information, it is able to classify and manage the input of measurement data. Optical illusions are good examples of such data management or distortion.

What do we know about the principle according to which biological data-processing systems function? They can hardly be compared to technical computer systems with similar functions. In the circuits of the nervous system no specially constructed storage units, amplifiers, or relays can be distinguished. Biological computer centers know only one kind of element: nerve cells, of which the human brain contains 100 million. The number of contact points between these nerve cells probably amounts to a billion. The number of possible connections between nerve cells is so great that details regarding the connection diagram of the brain will always remain a mystery. In the possibilities for combinations between nerve cells and their mutual chemical or electromechanical influenceability lies the secret of their suitability as elements for data processing. Signals can be amplified, analyzed, combined, and purposefully transformed by the appropriate circuits. The most important thought and control processes are still waiting to be explained. This applies in particular to the molecular foundations of data storage.

This brief excursion into the world of the nervous system is meant to convey an impression of the heights to which electrochemical processes have taken life, and of the importance of data processing for the mastery of the physical environment.

Electrical hunting and orientation

The waters of the lower Amazon are yellow with loam, which the river carries down to the sea. In times of flooding, remains of vegetation in enormous quantities drift downriver, creating rather difficult survival conditions for the fish. Since they cannot rely much on their eyesight, these fish have highly developed olfactory, tactile, and acoustic senses. The one method of direction finding and hunting that has proved to be particularly useful here is based on the emission of electrical impulses.

The electric eel of the Amazon (*Electrophorus electricus*), which grows to a length of up to 2 meters, is able to emit brief impulses of 500–800 volts and about one ampere. This is sufficient electrical energy to make a light bulb glow—and, of course, enough to make prey numb. In addition, the electric eel continuously emits brief electrical impulses of low voltage. Similar weakly pulsating electrical fields are also produced by numerous South American knifefishes (*Rhamphichthyidae, Sternarchidae, Gymnotidae*), of which there exist some 100 species. These electrical fields presumably serve in the detection of prey and obstacles, but also as means of communication and territorial delimitation. Numerous inhabitants of murky African waters also have electrical organs—among them the gymnarchids (*Mormyridiformes*), which occur in some 150 species, and the electric catfish.

In the waters of the other continents no true electric fishes are found, though many species living there are related to them. There are, however, many electric fishes in the ocean. Some 30–40 species each are known of electric rays (*Torpedinidae*) and electric skates (*Rajidae*). Then there is the stargazer (*Uranoscopidae*) and the sea lamprey (*Petromyzonidae*). Structures of electrical organs have even been found with some certainty in fish fossils about 300 million years old. Eleven different families of fishes are known to include species with electrical organs. This is an indication that electrical organs have developed repeatedly and independently.

Fishes that can produce strong electrical surges at call are found both in fresh water and in the ocean. The freshwater dwellers produce higher voltages (electric eel: 500–800 volts; electric catfish: 450 volts) than ocean dwellers (electric ray: up to 50 volts). This makes sense, since fresh water is a considerably poorer conductor. More thorough research has shown the electrical system of fishes to be constructed so as to deliver to the water the highest possible electrical power (current × voltage).

Most of the electric fishes that continuously emit low-voltage signals live in murky fresh water. The electric organ is composed of a large number of electroplaques (electrical cells), connected in series, and cell columns, connected in parallel, which have evolved from muscle cells and muscle bundles. In the electric eel, the electrical organ accounts for as much as 58 percent of the total body weight. It is subdivided into some 70 arrays side by side along the body, each one containing some 6,000–8,000 electrical cells wired in parallel like batteries. At a nerve signal from the brain, all the electrical cells are discharged within 0.001–0.002 second. In order for all cells to fire simultaneously, delaying mechanisms in the form of different conducting speeds are built into the trigger nerves. The anatomical fine structure of the individual electroplaques is somewhat different in its details for each species of electric fish, but the underlying principle is exactly the same, as described in the following with respect to the electric eel.

The electrical cell has a flat, waferlike shape. Its membrane is electrically charged like a nerve membrane. The charging occurs as in the nerve membrane—essentially, through the transport of sodium and potassium ions, which creates a negative charge surplus in the interior of the cell. Nerve ends are connected to only one flat side of the electrical cell. The other side is only connected to blood vessels and cannot be electrically excited. A nerve impulse reverses the polarity

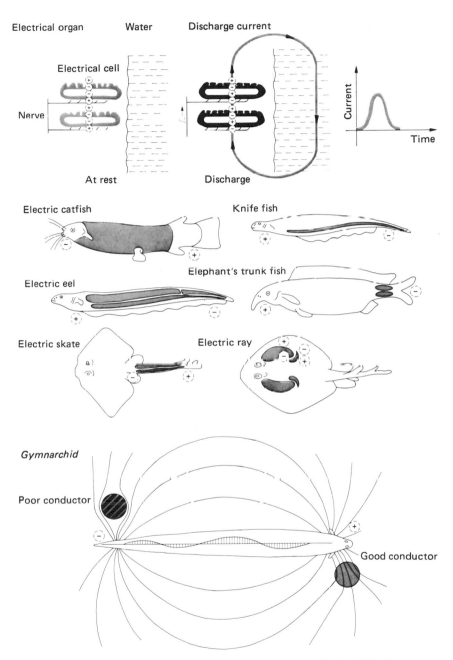

(Top) The discharge mechanism for a battery of electrical cells. (Middle) The distribution of the electrical organs in various fishes. (Bottom) The mechanism of direction finding by electrical field.

of the electrically excitable part of the membrane and thereby triggers a discharge of the entire cell. Since the electrical potentials of the opposite parts of the membrane are thus added to each other, the discharge voltage of a single cell exceeds the value of the rest potential and reaches about 0.1 volt. Since many electrical cells are connected in series, their individual voltages accumulate to the total discharge voltage of the electric fish.

If the fish uses this electrical energy merely to numb its prey, it has no need for any further electrophysiological devices. The matter, however, becomes much more complicated when electrical impulses are to be used to locate and recognize objects. For this purpose, low-voltage impulses are produced—either in a steady rhythm or else rather irregularly, depending on the species. *Gymnarchus niloticus* "broadcasts" day and night, without interruption, at a set frequency of about 300 impulses per second. Several species of South American electric eels emit super-rapid sequences of 1,100 or 1,600 impulses per second. Others use variable frequencies.

Whenever a difference of voltage builds up at two opposite poles of the electrical organ, lines of electrical flux penetrate into the watery surroundings of the fish and produce in it ionic currents. The distribution of the lines of flux depends in a very characteristic way on the conductivity of the surroundings. Good conductors concentrate the lines of electrical flux and thereby the electrical current; poor conductors repel the lines of flux and conduct correspondingly little current. Thus, when an object is introduced into the water and the conductivity of the object differs from that of the water, it will forcibly produce distortions in the course of the lines of electrical flux.

Lissman's experiments indicated that electric fishes are able to perceive such distortions. Gymnarchids learned to distinguish between bait into which metal and pieces of glass had been imbedded. They were able to perceive voltage gradients of 3×10^{-10} volt per centimeter, and their sensitivity threshold for current impulses was estimated at 3×10^{-15} ampere. Since the electrical background noise is considerably stronger, we must conclude that many thousands of receptor cells function together and compose a median value of the signal.

Strangely, it has not yet been possible to find a noticeable and sharply defined sensory organ that would record intensity changes of the electrical field. But it has been shown in *Gymnotidae* that the lateral line organ (an agglomeration of sensory cells along the body of the fish) is sensitive to electrical fields. Similar field receptors, distributed in fixed patterns over the body, were found in the African elephant nose (*Mormyridae*). These fishes have similar body shapes and therefore similar mechanisms of motion. They keep the body almost rigid and swim with the greatly elongated dorsal and end fin (respectively), over which they can direct propelling waves in both directions for swimming either forward or backward. In this way they can dispense with slithering, which would make it more difficult to determine the exact direction of field changes. This is another example of the strong influence the application of certain physical mechanisms had on the evolution of the animals.

No artificial direction-finding systems like those of electric fishes have been developed, and in all probability none ever will be. If the main purpose of the electrical activity of fishes is indeed the discovery of objects of different conductivity on the basis of shifts in the lines of electrical flux, then the question arises whether the evolution of such a complicated organ was altogether worthwhile. Water (including fresh water) is such a good conductor that the electrical field drops rapidly at a distance from the fish, and objects farther away than one or two fish lengths can hardly be perceived. Thus, an electric fish on a nocturnal hunting expedition can at best scan its immediate surroundings by electronavigation. Another question is, of course, whether a prospective victim will hold still and allow itself to be scanned. Neither does it seem that the ability to distinguish objects by their conductivity would offer much of an advantage in the habitat of electric fishes. In the lower Amazon you will look in vain for nonconducting stones. The water-soaked muddy floor certainly does not provide a good electrical contrast. The exuberant profusion of aquatic plants through which an electric fish must maneuver is, with its increased salinity, probably close in conductivity to the bodies of animals. Other weakly electric fish live in torrential rapids in shallow water. In such circumstances, the envelope of lines of electrical flux, which fluctuates greatly with the depth of the river, can hardly be expected to allow for reliable electrical location of prey. Is electronavigation a technical aberration, or does it harbor other secrets?

It is strange that the electrical sense has never been viewed from the electrochemical angle. Actually, direction finding by shifts in the lines of electrical flux as it is being discussed these days would function just as well in a vacuum or in air. Electric fishes, on the other hand, trigger currents in the conductive water. The currents hit whatever objects happen to be present in their surroundings and, given the proper conductivity, penetrate them. But currents imply the migration of ions. Wherever they penetrate a surface, electrochemical reactions may occur. Ions give off charges when they themselves cannot penetrate a surface, and by doing so they often modify their stability and react chemically. When the voltage impulses are brief, electrical charge transfers occur on a surface. Meanwhile, absorbed ions and molecules may detach themselves from the surface. Therefore, currents, even if they are very small, have a destructive and destabilizing effect. A marine creature penetrated by a current would have to lose a corresponding quantity of organic ions to the surroundings. It is easy to figure out that currents of less than a millionth of an ampere would still release considerably more organic ions when hitting prey than a fish with a sense of smell needs for comfortable perception.

These speculations lead to the interesting possibility that weakly electric fishes raise the scent level of their environment electrochemically through their electrical impulses. What advantages would a fish derive from such a hunting technique? No doubt the effective range of electrochemical contact would be much larger than that of direction finding by electrical field. The emission of brief electrical impulses would make sense insofar as organic tissues are good conductors only for a brief span, after which their conductivity abates rapidly. The given conditions for a blind chase guided by the olfactory sense are nothing short of fabulous: The hunt along an electrochemical scent spoor would not be hindered by aquatic plants or by rocky obstacles in rapids, which could be detected at close range by electrical direction finding. Since electrical impulses are periodic, a fleeing marine animal would also leave periodic scent waves, whereby the ability to detect scent molecules of the located prey would be extraordinarily increased. Such molecules would be easily distinguishable from the constant scent level. On the basis of the time sequences of olfactory signals, the electric fish in pursuit would be in a position to comfortably establish the speed of the victim, and (by adjusting its own emission frequency) also the distance.

This hypothesis could, of course, be confirmed or refuted only by specific experiments. For our purposes it is enough to note that a closer examination of the electrochemistry of periodic current impulses has led to a new and supplementary concept of the direction-finding mechanisms of electric fishes. It seems a good idea to start from the premise that biological mechanisms ought not only to be elegant, but also to work efficiently.

9

Physicochemical Methods and Tricks

Dancing on water

The surface of the water is a world in itself, ruled by very special physical laws. This world too was turned by nature into a habitat of life. How elegantly this was managed becomes obvious when we watch the effortless cavorting of water skippers (*Geris*). They move along by fits and starts, sometimes even by jumps, gliding on their hind legs like on runners over the surface. The middle legs serve as rudders, while the front legs are used only to grasp prey. Looking closer, we notice that the water is indented by the weight of the insect where the legs touch down. The water seems covered by a thin film which carries the water skipper. But there is no film; what brings about the surface forces is simply the fact that the water ends there.

Water molecules exert mutually attracting forces on each other. At the proper temperature they also cause water vapor to condense into droplets of mist. Under water, a molecule is exposed to the same forces from all directions; but when the molecule is located on the surface, these forces are one-sided and they try to draw the molecule under. The water surface counters by assuming the shape that is energetically most favorable. An attempt at penetrating or reshaping the surface involves work; tension must be overcome. It is this surface tension that the water skipper exploits for its locomotion. In order to do so effectively, it must comply with very specific prerequisites. The buoyancy of the water does not depend only on the weight and the shape of the swimming insect. The same body, if it is not too heavy, may either float upon the surface or drown in the water.

What is decisive for the ability to swim is the interaction between the liquid and the swimming body. Its extent determines the angle of contact between the water surface, the air, and the body surface.

The weight of the water skipper indents the surface. On the bottom of the water, the refracted light shows the points where the legs touch the water.

For optimal buoyancy of the water, this angle must be as large as possible; that is, the floating object must be water-repellent. If the object were moistened by the water, the water level would quickly close in upon it and it would sink. The water skipper is ideally equipped for this wet pavement. Even the underside of its body, which normally does not come in contact with the water, is covered with a fine, water-repellent coat of hair. One could play a mean trick on this creature by adding to the water some detergent, which makes a surface wettable. The insect would be helplessly sucked under.

The surface tension of water presents a serious problem for insects and many other small creatures when they want to drink. Depending on the surface properties of their oral organs, they run a bigger or smaller risk of being caught in the water or not being able to penetrate the surface film. They adjust their way of life accordingly and obtain moisture from their food, from very small droplets, or from moist soil. Water-dwelling insects that come to the surface to breathe do not have these options. Sometimes such an insect must resort to eccentric maneuvers to pry its breathing hole, which is surrounded by water-repellent hair, loose from the water surface. The larvae of mosquitos use for this purpose a specific flipping motion. The wormlike *Limnophila* larva must twist its body out of the water and roll it over its own air hole in order to push the latter below water.

Taking strolls on the underside of the water surface is a favorite hobby of the beetle *Octhebius*, which in the course of this exercise hangs by its water-repellent "tiptoes."

Getting ahead without moving a muscle

The large-eyed rove beetles of the genus *Stenus* make use of surface tension for their locomotion. Driven by an invisible force, they glide over the water without any body movement. The secret of such "surfactant swimming" lies in glands at the end of the abdomen which excrete an organic substance to lower the surface tension of the water. While the outpouring organic molecules cause the surface tension of the water to collapse in circles around the end of the beetle's body, the effect of the original, higher surface tension on the tarsi of the insect is maintained. This asymmetrical distribution of forces acts to pull the animal forward.

The principle used in this motor mechanism can be reproduced very simply: If talcum or a similar inert powder is sprinkled on the water surface and a drop of oil is added, the spreading oil sweeps the powder before it like a plough. The propelling force is located where

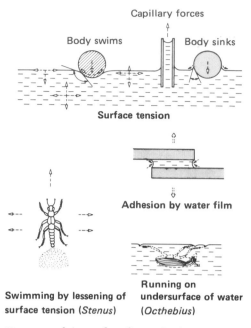

Capillary forces

Body swims   Body sinks

Surface tension

Adhesion by water film

Swimming by lessening of   Running on
surface tension (*Stenus*)   undersurface of water
    (*Octhebius*)

Nature exploits surface forces for locomotion and adhesion.

the oil starts to overlay the water. It is the result of molecular movement released by surface forces, and it produces speeds of 20 centimeters per second. A graphic example is presented by a piece of camphor placed on a clean water surface. As a consequence of the uneven dissolution on its surface, it may dash about in rapid and eccentric moves.

The rove beetle's chemical drive demolishes the tension of the water surface, to the danger of the water skipper. But here too we find confirmation for the farsightedness of evolutionary development: The rove beetle is not a true water insect; it lives on the shore. When it gets into the water by accident, it tries to get out as fast as it can.

The explosive reactor of the bombardier beetle

The most spectacular chemical reactions are explosions. An explosion runs its course in a fraction of a second. Whatever heat cannot be given off rapidly imparts additional acceleration to the reaction. When gases are formed as products of the reaction, high pressures are created which burst the reaction chamber or release a high-speed stream of gas through a jet. One would guess that such violent reactions are incompatible with the delicate world of organic molecules and structures. Nevertheless, in at least one case nature has experimented with an explosive reactor and developed it to a remarkable extent.

The pygidial glands of the bombardier beetle (subfamily *Brachyninae*) produce a mixture of two hydroquinone compounds and 23 percent hydrogen peroxide in two separate chambers closed off by muscular valves. When the beetle is excited, the valves of the storage chambers open and their contents flow into a reaction chamber with thick walls. At the same time, an enzyme is added which acts as a catalyst and produces an explosive reaction in which hydrogen peroxide is decomposed and oxygen is released. The gas pressure forces the caustic quinone solution out of the reactor and shoots it at the aggressor.

The explosive reactor of the bombardier beetle is remarkable not only because it has the essential elements of a rocket engine, but also because the same catalytic decomposition of hydrogen peroxide played an outstanding role in the early stages of reaction engines. The hydrogen peroxide engine was used during the second world war and thereafter as a starter rocket for the V-1 "buzz bomb" and for manned aircraft, to drive the fuel pumps of the V-2 rocket, and to propel torpedos.

Adhesion

Many people are annoyed when an exciting situation causes the palms of their hands to get moist. They should instead welcome the farsighted operation of their organism, which means to get them ready for events with which they might have to "come to grips." Moist palms offer the same advantage a lumberjack obtains by spitting into his hands before gripping his axe. Adhesion is produced by the interplay of intermolecular forces (the so-called van der Waals forces) between the electron shells of molecules. These forces are too weak to effect a normal chemical bond, but their strength is quite adequate to make a solid body adhere to another, or a liquid film to a solid surface. It is adhesion that makes chalk stick to a blackboard, or water drops to a window pane.

Adhesion is produced by the same kind of forces that are responsible for surface tension. Moreover, surface tension is participating directly in the mechanism of adhesion when fluids are bonded. The forces of adhesion may appear relatively weak to us humans, but they can be important to a small creature that has a large surface relative to its weight. Bees and wasps are able to run across a perfectly smooth verticle glass pane because their soles are made for adhesion. Some tropical tree frogs use, besides their suction-cup toes, the adhesiveness of their broad belly surface to land securely on a very narrowly limited smooth spot after a daring jump. Adhesion also plays an important part in the function of suction equipment, and it gives the rim of a suction cup pressed to a surface its first hold.

Adhesive surfaces in nature are always moistened by the secretion of a liquid. This improves adhesion considerably. Without a liquid film there would be few points of contact between two solid bodies because of surface roughness. A liquid creates ideal contact by moistening the surfaces, and it gives two bodies additional hold through its surface tension. This trick of nature is very easy to copy and to understand if you press first a dry and then a moist piece of paper against a window pane. Good contact surfaces made adhesive with water can hold weights of up to half a gram per square millimeter. More viscous liquids (that is, liquids whose molecules are more difficult to separate) improve adhesion further. No wonder many small animals run or creep through life on moist soles.

The only difference between an adhesive connection and a connection by glue is that with glue the intermolecular forces of the liquid molecules are strengthened by chemical or physical anchoring or by processes that occur during hardening. Glued connections are very common in nature; they are used for the anchoring of eggs or cocoons, the catching of insects, the transporting of seed, and the putting together of complicated structures of sand grains or plant material.

The transpiration power stations of plants

A suction device will pump water upward to a height of 10 meters. A water column 10 meters high creates exactly one atmosphere of pressure. Even if you create a vacuum above the water, you cannot raise it higher; the weight of the water does not allow it.

One need not stand before a giant California redwood to understand that the water pumps of plants are extraordinary. And it has been clearly demonstrated that the energy used to transport water in plants is not generated by chemical mechanisms. A tree sucks up to its highest branches even poisonous solutions which are sure to kill off its living cells. An oak can pump at a rate of over 43 meters per hour. The same speed is attained by wheat. Many other plants, though, transport water at a more leisurely pace. The average birch transports on a sunny day 100–200 liters of water, while one liter is pumped in the same time span by a single sunflower.

A simple calculation based on these figures allows us to estimate the consumption of energy during the pumping process. Let us assume that one square centimeter of a trunk pumps during one day one liter of water to a height of 100 meters, which is a very conservative assumption and by no means the maximum attainable. Then 2.72 kilowatt-hours of energy would be consumed per day on one square meter of trunk. That is enough to keep eleven 100-watt light bulbs burning all day long.

Experiments have proved that the energy used in pumping is produced by the evaporation of water in the leaves. Whenever the evaporation of water in the leaves increases, the pumping action in the stem of the plant also increases. But how did nature manage to obtain mechanical energy from the process of evaporation? The idea is simple and ingenious. During the process of evaporation, water molecules leave the water because of their kinetic energy and pass into the air. Since fast molecules penetrate the surface of the water more easily than slow ones, heat is extracted from the remaining water. (This is why we feel a pleasant coolness when we wet our body on a warm day.) If it were possible by some added force to retain the evaporating water molecules, with their high kinetic energy, they would lose some of their energy but would transmit it to the remaining water. They would therefore be able to pull up the water beneath them. And it is in fact possible to perform this feat by letting water evaporate through thin tubes—capillaries—as plants do. In capillaries, the surface of the water that is held fast by surface tension is large in comparison with the volume it encloses. Therefore, a water molecule would have to work harder to tear loose from the tip of a capillary, since it is held tighter by surface tension. More simply, it would get out only by pulling up another water molecule to take its original place. Thus, water in thin capillaries evaporates less, but pulls up the water beneath. Besides, a very thin water column in a capillary would hardly break up under its own weight, since it is supported by intermolecular forces on the surface of the column. It has been calculated that such a column could ideally attain a height of more than 3,000 meters without losing cohesion. Plants pump water through microscopically small tubes which reach from the roots to the tips of the leaves without interruption of the water column. Energy is supplied by the process of transpiration, which can only occur when enough mechanical energy is released to keep the water supply going. In other words, plants operate power stations that convert heat into mechanical energy.

Nature did not refrain from using energy produced in the power stations of plants for other important functions of life. One example is the desalination of seawater by evaporation energy. At the edge of shallow coastal waters of tropical seas we find the luscious green of mangrove swamps. Mangroves can live on the saline water of the ocean, which destroys other green terrestrial plants. In some species of mangroves the sap is almost salt-free, though the roots are washed by sea water. They extract the salt by using the transpiration energy in the narrow capillaries of their roots to suck up the sea water and then filtering it through thin membranes in which the salt is detained.

A similar procedure in which sea water is filtered through membranes under pressure has been in use for some time in man-made desalination installations in arid zones. Unfortunately, these installations use up a lot of energy, whereas green plants derive cheap energy from the sun.

Antifreeze

Insects have no mechanism for regulating body temperature. Nevertheless, numerous insects winter outdoors in the rough climate of North America and northern Asia, withstanding temperatures below −20°C. The Japanese caterpillar *Cnidocampa*, which spends the winter in a watertight cocoon, can be kept for many weeks at −15°C and even cooled to −20°C without suffering any damage.

If tropical insects were to be exposed to such temperatures, they would die instantly. Research has established that insects that can withstand cold in winter are not able to endure equally low temperatures in summer. The obvious conclusion is that they produce some protective agent against the cold when winter approaches.

Antifreeze works by lowering the vapor pressure above a watery solution. The freezing point of the mixture is thereby lowered, since it lies at the temperature at which the vapor pressure above ice is equal to that above the mixture. In order to obtain a sizable lowering of the freezing point, as many molecules as possible of the antifreeze must be dissolved in the water. It is therefore necessary that the antifreeze mix well with the water, that it have a high boiling point, and that its molecules be small so that many will be contained in a specific volume of water. Glycol has proved very effective in radiators of motor vehicles. By mixing 50 percent water and 50 percent glycol, one lowers the freezing point of the mixture to −34°C.

Research on wintering insects has shown that their blood contains glycerol, which in its structure and properties differs only slightly from glycol. In the caterpillar of a North American silk moth (*Hyalophora cecropia*). 3 percent glycerol was found in the blood. The wintering larva of the parasite *Bracon cephi* contains as much as 20 percent glycerol in the blood. Proof was found that fish, too, can sustain life below 0°C. Fish live in the waters of northern Labrador and Antarctica at −2°C. It has been suggested that glycoproteids (compounds of proteins and carbohydrates) function as antifreeze in fish.

The antifreeze in the body fluids of insects and fishes lower the freezing point, no doubt. But this alone does not explain how some insects are able to survive extremely low temperatures; the concentrations of these compounds are simply too low. There must be an additional device that prevents the water in their bodies from freezing. The only alternative to the lowering of the freezing point by antifreeze is offered by controlled undercooling. Ice crystals need a nucleus of crystallization in order to start forming. This seed (which may be a grain of dust or a large molecule) gathers the first water molecules into ice-crystal structures through the arrangement of its surface atoms. Only thereafter are the rest of the water molecules able to build up in rapid succession around this tiny ice crystal nucleus and go on to form large ice crystals. When absolutely pure water is cooled, the lack of crystallizing nuclei allows for considerable undercooling without ice being formed. Water can easily be cooled to −10°C, but if one tiny ice crystal is thrown in, then there is suddenly nothing but ice. Nature may have discovered a method of eliminating crystallization nuclei from water, or of checking them. A mechanism of this kind is not known in physiochemistry, but the adaptation of insects to the cold should be an inspiration to look for one.

When water is scarce

The most important precondition for life in the desert is the water supply. Plants, and animals too, dig deep wells. The roots of desert plants often reach 10−15 meters into the ground for water. When the Suez Canal was built, the roots of a tamarisk were found at a depth of 30 meters. Water is sucked in through the osmotic pressure of the root cells, which allows water molecules from the soil (which has a slightly lower concentration of salts) to penetrate into the highly concentrated saps of plant cells. In desert plants, osmotic pressures of up to 100 atmospheres have been measured—pressures that could raise a water column 1,000 meters high.

Widespread root systems enable many desert plants to specialize in collecting and storing water during the short, floodlike thunderstorms of the rainy season. Many different storage methods have been developed for this purpose. Water reserves are kept in underground tubers, in bulbous stems, or in thickened leaves. The water is pumped into the storage spaces by osmotic pressure, or is sucked up through thin capillaries. In addition, nature has provided swelling substances which store the water between their giant molecules and swell like moist loam.

Once the water has been collected, the struggle begins against loss of water through evaporation. The water must be transported close to the surface of the plant, where sunlight can split it through the photochemical reaction of chlorophyll. To prevent excessive evaporation at the plant surface, nature has applied a number of interesting devices: The ratio between plant surface and plant volume is drastically reduced. Often leaves are entirely eliminated and the photosynthesis occurs on the surface of the stem. A prime example is the cylindrical torch thistle (cereus). The smallest possible ratio between surface and volume has been attained in the spherical cacti. It is many hundreds (even thousands) of times smaller than those of plants that live in moist zones. Other desert plants simply drop their leaves during periods of intense heat, or periodically roll or fold their leaves to control solar irradiation and the discharge of moisture. Frequently, plant surfaces are hardened and additionally protected against water loss by a layer of wax. Another clever device is the dense hair felt which numerous plants of arid zones wear on their surfaces to reduce the loss of moisture to the wind. This gives the same protective effect that the Bedouins achieve with woolen cloth. Some plants unite into dense pads and retain moisture and temperature by creating a collective temperate microclimate.

Survival in the desert is rougher on animals and humans, because they are much more dependent on a regular water supply than are plants. The history of the deserts is rich in stories like that of the caravan of 2,000 man and 1,800 camels that perished of thirst in the Sahara in 1805 because one water hole had dried up. But in well-adjusted desert animals nature has succeeded to an astonishing degree in rationing the water consumption. They spend as many as possible of the hot hours of the day in their narrow subterranean burrows, where the atmospheric moisture is high. Camels at rest turn with the sun and always expose to it the narrow side of their body. When many camels are gathered, they crowd together to provide each other with shade and reduce water evaporation. The urine of desert animals is highly concentrated, and their droppings are dry. Through the intake of nourishment and the metabolic process, so much water can be obtained under favorable conditions that some desert dwellers are independent of free drinking water. Kangaroo rats manage by this method, though they feed mainly on dry seeds.

Certain desert termites build galleries that reach as deep as 40 meters into the sandy soil. These well ducts are masterpieces of underground engineering, since their walls must be carefully cemented to secure them against collapse.

Desert animals deal with the difficult problem of losing moisture through exhaling by breathing more slowly. An interesting physical device for the recovery of respiratory moisture was recently discovered: The temperature of air exhaled by kangaroo rats was found to be only 25°C, while their body temperature is about 38°C. Evidently, the exhaled air is cooled off by heat-exchange installations in the respiratory tract. Thus, moisture can be extracted from the air still inside the body by condensation of water at lower temperatures.

Marine mammals and seabirds, too, had to overcome difficult problems of water supply. Whales and seals, which live on fish, have less difficulty. The concentration of salt in fish is considerably lower than that in sea water. Sea mammals that feed on other marine animals excrete highly concentrated urine and thereby accumulate in their body desalinated water. Gulls, pelicans, cormorants, and albatrosses excrete excess salt through salt glands on the head. The sea iguana of the Galapagos Islands, the penguins, and the sea turtles likewise desalinate water in appropriate glands. Nature has thus found an economical method for desalinating sea water. Not much is known to us about the process apart from the fact that it takes place in the thin biological membranes and is powered by the same molecule (ATP) that moves the muscles and supplies many other vital processes with energy.

The Atacama Desert of northern Chile is one of the driest desert regions on earth. In some areas it has not rained within the memory of man; in others it rains only once every few years. On the average, no more than 2 centimeters of water per year reach the bone-dry desert soil. In spite of this acute lack of water, the hills of mountain ranges near the coast harbor some vegetation. It is not very dense, but it includes many large cacti and the stately tamarugo tree.

These desert plants do not find a trace of moisture in the soil. They extract whatever moisture they get from the fog that frequently forms when the moist sea air cools. Wandering at dawn through these fog-clouded zones of vegetation, one can easily make some interesting physical observations. Large drops of water adhere everywhere to the plants. A closer look shows that they almost always hang from the tips of the long sharp thorns.

Some basic concepts of electrostatics and surface physics suggest why tiny particles of fog collect at the tips of the thorns. The ground and its plants have a negative charge relative to the air, and every

For a whole day I searched the Valle della Luna (moon valley), north of the oasis San Pedro de Atacama in the desert of northern Chile, a valley that consists of nothing but salt, sand, and gypsum, and which I was convinced could not harbor a trace of life, . . .

. . . and then I found this small, delicate plant on a dune. To me it seemed a miracle that it could exist there. Its survival weapons are sophisticated.

charged object creates a charge separation through electrical influence in an uncharged object—such as the fog droplets may be. The charge separation is aligned in such a way that opposite charges are attracted. When a small fog droplet approaches the tip of a thorn or a plant hair, the electrical field force there is so strong that a charge can pass over into the droplet. These surplus charges, because of their mutual repulsion, considerably lower the surface tension holding the droplet together. Through this so-called electrocapillary effect, the droplet loses its stability, flows onto the thorn, and spreads over its surface.

Other undemanding desert plants obtain water by a clever choice of location. The green of plant life sometimes shimmers through the Atacama Desert's many pale, translucent quartz pebbles. These plants obtain filtered light through the mineral, which evidently is their method of finding protection against excessive heat and which permits them to collect the condensation water forming on the cool underside of their protective shield.

10

Mathematical Laws in Biological Structures

Mathematical games in nature

For someone who knows the rules, it is not particularly difficult to construct a pentagon with a compass and a ruler. But behind this routine geometrical procedure there are complicated mathematical laws. Nature, having created pentangular symmetry in innumerable starfish, sand stars, and other echinoderms, must have preprogrammed these mathematical correlations on the molecular level. (A few starfish and sand stars have broken out of pentangular geometry and developed perfect hexagonal or heptagonal shapes, and in a few isolated cases starfish with 15–20 tentacles have been found.) Almost all other multicellular creatures show two-sided symmetry.

The nautilus is a pleasure to look at because of the perfectly fashioned spiral form of its shell. Snail shells too are spiraled, though the spirals seldom lie in one plane. Even the array of seeds in plants or the distribution of petals frequently occurs in spiral form. Viewed with regard to their mathematical properties, spirals are among the most elegant geometrical shapes. This is true in particular for the logarithmic spiral, which bisects all straight lines originating from its center at the same angle. This spiral occurs in nature. Whether this spiral is differentiated, integrated, raised to a higher power, or its root extracted, the result every time is another logarithmic spiral.

In many plants the leaves are arranged in a spiral along a branch. Banana plants (*Musaceae*) have only left-threaded spirals; all other plants seem to have right- or left-threaded spirals with equal frequency. It has been known for a long time that spiral arrangements in nature obey a mathematical law: Pick a leaf and count all the leaves along the axis of the stem up to the one that is exactly above the first leaf. You will get a characteristic number. By dividing this number by the

A very few species of starfish have broken out of the pentangular rule and are six-armed (*Meyenaster gelatinosus*, Chile).

number of spiral turns between these two leaves, you get a fraction that is typical for the particular plant. The interesting thing is that all leaf positions found in nature add up to a series of fractions: 1/2, 1/3, 2/5, 3/8, 5/13, 8/21, 13/31, . . . The progression starts with 1/2 (adversifoliate position) and continues in such a way that each fraction contains the sum of the numerator and the denominator of the two preceding fractions. This is the Fibonacci progression (named after an Italian mathematician), and it tends to a limit of 0.382. Between this limit and the golden section (the familiar mathematical quantity $(5 - 1)^{1/2}/2 = 0.618$) there is a remarkable correlation: $0.382 + 0.618 = 1$. These mathematical relationships cannot be the result of chance. They must be based on molecular mechanisms we have not yet discovered. Neither do we know why some plants grow into pretty spirals. Beans, gourds, and cucumbers always grow in right-threaded spirals; hops and honeysuckle (*Caprifoliaceae*) prefer left-threaded spirals. The direction of the thread is genetically determined and cannot be changed by experimental intervention.

Nature contains many examples of mathematical games. We have no clue to the formulas according to which mathematics is applied in biology, and sometimes the purpose of such application is equally unclear. However, it certainly seems that life is fluent in mathematics, the language of physics.

Biological clocks

Time is one of the most important physical quantities. A precise knowledge of time is necessary for the coordination of various processes. It is therefore not surprising that living creatures have inner clocks. The sense of time in animals has been demonstrated by innumerable experiments. Even people who are far removed from a natural way of life can train themselves to wake up at a set time without an alarm clock.

The most impressive proof of animal chronometry is offered by navigating birds. A bird released far from its home base often needs no more than one look at the sky to determine the direction of its return flight. Such direction finding would be possible only with a precise "clock."

We do not know how living organisms measure time. However, we can make quite a bit of headway simply by asking how chronometry works, because then we discover that periodically occurring processes are an indispensable prerequisite for the measuring of time. Time can be measured only if it can be counted by the "pendulum swings" of a regularly recurring state. The mechanical ticking of a clock, the oscillation of an electrical circuit, or the periodic flipping of a molecule are milestones of time. Are there such periodic processes in organisms?

Periodic mechanisms are unequivocally defined by physical properties and mathematical equations, and always include a process aimed at reestablishing the original state of the oscillating system. For a mechanical coil spring this process is the spring constant, which builds up elastic energy and causes the spring to snap back. For a pendulum it is the gravity of the earth, which must be counteracted by mechanical work until the kinetic energy has been consumed. In animated organisms no pendulum devices have so far been discovered. Instead, innumerable periodic processes have been observed which are either mechanical or electrical or chemical in nature. The most striking periodic process in the body is the heartbeat. Besides this, there exist many nerve cells which transmit periodical electric impulses. The flying apparatus in many insects is driven by a periodic mechanism. Periodic processes occur on many membranes. They are especially pronounced in the mitochondria (the small power plants of the body cells, in which the energy-rich molecule ATP is assembled). Numerous chemical reactions in which enzymes participate show a pulsating course.

All these rhythmic biological phenomena can be traced back to the fact that certain chemical reactions can oscillate periodically like clockwork. When certain chemical substances are poured together, such

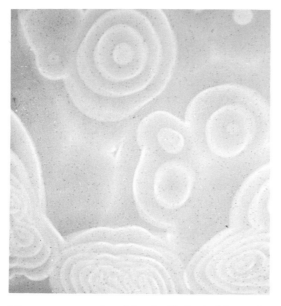

The patterns produced by the Zhabotinskii reaction

periodic chemical reactions can even be observed in a glass vessel in the easily produced Zhabotinskii reaction. It is arrived at as follows: 5 grams $NaBrO_3$ (sodium bromate), 2 cubic centimeters concentrated sulfuric acid, and 67 cm^3 water are mixed with 11.65 cm^3 malonic acid solution (10 g malonic acid per 100 cm^3 water). Then 5.83 cm^3 NaBr solution (10 g NaBr per 100 cm^3 water) are added, drop by drop, at 0°C. Then 0.78 cm^3 of 0.025 molal ferroin solution are added with a pipet to 20 cm^3 of the solution produced by the previous reaction, and the end product is thoroughly mixed. For one hour— until the energy-supplying partners of the reaction have been used up—waves pulsate through the liquid. Beautifully colored concentric rings spread slowly outward; spirals wind upward or downward; waves overlap and either extinguish or reinforce each other. The Zhabotinskii reaction is very complex. We understand it so far only in rough outlines, and a computer is needed to prove that its essential features can indeed be described. It is the autocatalytic chemical reaction steps that this reaction has in common with all other periodic chemical and electro-chemical reactions. A reaction is autocatalytic if, for instance, the original product is doubled in the end result. This means that the reaction accelerates itself. But in a similar way a reaction can also slow itself down.

The simplest autocatalytic reaction that can lead to oscillation is the Lotka-Volterra reaction. It describes the interaction between predators

Ducks on their journey (coast of Chile).

and their prey. Foxes get into an area inhabited by hares. They proliferate quickly and gradually cut down the hare population. In this way they bring about their own starvation. Then hares proliferate again and the game starts all over. The same process can be written in a chemical formula: $A + B \rightarrow 2B$, $B + C \rightarrow 2C$, $C \rightarrow D$. Like all oscillating systems, this one is characterized by a feedback mechanism (a mechanism that can alternately accelerate the system and slow it down). In the last analysis, all periodic processes in living organisms must be traceable to such chemical reactions with feedback. They also drive the clocks by which life measures time.

Orientation and navigation

A search for direction in an unfamiliar and structureless environment is possible only with the help of astronomical or geophysical factors. The magnetic field of the Earth is one suitable geophysical quantity. The positions of the stars or the sun offer astronomical points of reference; but since the Earth is rotating about its own axis and around the sun, the sun and the stars wander across our skies in orbits that continuously shift with the seasons. Therefore, a simple and practical way to determine the geographical latitude of a position is to choose specific fixed points. The pole star presents itself for this purpose. It points with reasonable precision in the northern direction of the Earth's axis, and its position does not change with time. During the day the best choice for establishing the southern direction is the apex of the sun's orbit at noon, which, if the season of the year is known, also serves to determine the geographical latitude. Geographical longitude is more difficult to determine. Owing to the Earth's rotation, all observers on the same parallel pass successively the same astronomical observation points. That is why the measurement of time is indispensable for the determination of geographical longitude.

The exact determination of a location is a rather complex procedure that requires a sound knowledge of physical correlations and precise measuring instruments. It seems therefore astounding that nature succeeded in equipping numerous animals with a well-functioning sense of navigation. The itineraries of some migratory birds are, even by the scale of our era of intercontinental traffic, overwhelming. The Arctic tern (*Sterna paradisaea*) flies annually from the Arctic to Antarctica, covering up to 29,000 kilometers. The golden plover migrates from the Canadian tundra to the southern pampas of Argentina, and the Alaskan members of the species (*Pluvialis dominica*) are drawn to Hawaii, Tahiti, and other islands of the Pacific. The slender-billed shearwater (*Puffinus tenuirostris*) circumnavigates the northern Pacific. European birds such as the white stork and the chimney swallow fly all the way to the south of Africa. The wandering and the royal albatrosses travel the world continuously, flying over the cold southern seas, except when they nest.

We know for certain that the navigational sense is inborn in some birds. A remarkable example is the young cuckoo, which starts out on its winter journey without ever having caught a glimpse of its parents. The young of the golden bronze cuckoo of New Zealand (*Chalcites lucidus*) fly annually several thousand kilometers over the ocean to Australia and the Salomon Islands, to which the parent birds have preceded them.

There is also convincing proof that some birds not only determine direction but actually navigate: Eighteen nesting Laysan albatrosses (*Diomedea immutabilis*) that were in the way of traffic at an air base were flown from Midway to distant coasts all around the Pacific. Fourteen of the birds returned. One of them covered the distance of 6,600 kilometers from the Philippines in 32 days. Five birds that had been exiled to an atoll 2,660 kilometers south of their usual territory were back within 12 days. Two albatrosses that had been released on the northern Pacific coast of the United States flew the distance of 5,100 kilometers to Midway in 10 and 12 days, respectively. One of these birds traveled a daily average of 510 kilometers, the other an average of 425 kilometers. They must have flown almost directly toward Midway. There is other proof of extraordinary navigational capabilities: A Manx shearwater (*Puffinus puffinus*) that had been taken from its nest on the coast of Wales and brought to Boston covered the distance of 5,440 kilometers to its home in 12$^{1/2}$ days.

Not only swift birds but also ponderous mammals undertake extensive migrations. Many whales travel regularly from their antarctic and arctic feeding grounds to subtropical and tropical seas, where their young are born. The Californian gray whale makes an annual journey of almost 10,000 kilometers from its breeding grounds on the west coast of Mexico to its feeding grounds in the Bering Sea, taking 80–90 days to cover the distance. Finbacks from the Bering Sea and blue whales from the north Pacific travel every winter more than 12,000 kilometers to the Indian Ocean.

Some turtles, too, cover distances that are very great relative to their size and swimming technique. Green turtles (*Chelonia mydas*) from the Brazilian coast swim 2,500 kilometers to Ascension Island, in the middle of the south Atlantic.

The monarch butterfly (*Danaus plexippus*) of Canada and the United States migrates from northern latitudes to winter in California, Mexico, and Florida.

The migration routes of terrestrial mammals are generally much shorter. This is not surprising, as locomotion over land is much more laborious and more easily blocked by natural obstacles. Nevertheless, some land creatures execute astounding migrations. The caribou of Alaska and northern Canada, for instance, travels up to 1,300 kilometers to its winter quarters.

Less spectacular navigational feats have been observed among a great many animals. Whether one is considering toads which wander annually in swarms to a common breeding swamp or insects which return to their home hive, their navigation is in all probability based on geophysical or astronomical laws.

Over the last 35 years, research into animal migration with the help of radar, by experiements in planetaria, and above all by systematic marking and observation of animals has brought forth a great abundance of interesting data. Surely the most significant step forward has been the experimental proof of astronomical navigation by some migratory birds. The premise for a series of very successful experiments was given by the observation that caged birds of passage, when they get a glimpse of the sun or the stars, crowd against the bars of the cage in the direction of their migration. By clever mirror tests and experiments in planetaria, G. Kramer and F. and E. Sauer established without a doubt that numerous migratory birds use either the sun or the fixed stars for navigational purposes. Even with this clever experimental method, the exact mechanism of astronomical navigation has not yet been decoded.

Some of the complications encountered are demonstrated by the example of the homing pigeon, which is valuable to science because of its navigational abilities and its trainability. The homing pigeon uses the position of the sun as a direction-finding aid. In addition, it uses its inner clock to measure time. If the bird's daily rhythms are altered, it makes mistakes in orientation. For each hour the rhythm is disturbed, the pigeon deviates another 15° from its direction (K. Schmid-Koenig). But the ability to determine direction by the position of the sun and a sense of time does not in itself suffice for navigation. The bird must also find out where it is. Very good homing pigeons can return home within a day from a distance of up to 1,000 kilometers; they must therefore be able to determine their position with exactitude. G. V. T. Matthews of Cambridge University is of the opinion that homing pigeons can read off the sun more than just direction. According to him, they perceive the motion of the sun across the sky and can thereby estimate the course of its path to the highest position at noon, which is a measure of the geographic latitude at the point of observation. Unfortunately, this hypothesis has not been borne out by corresponding experiments. It seems after all that homing pigeons can determine only direction with the help of the sun.

For more than 100 years the view has been repeatedly advocated that homing pigeons and other animals use the magnetic field of the Earth as an aid in direction finding. This idea suggests itself technically, but it is rather problematic from the physiological point of view. Nobody can imagine what a magnetic organ that would react to the weak magnetic field of the Earth would look like. Repeated attempts have been made to impede the navigational capacity of homing pigeons by having them carry small magnets. For a long time all such experiments failed, but the experiments of C. Walcott and W. T. Keeton brought a surprise. They traced the previous results to the fact that the homing pigeons had usually been sent out in good weather and could therefore orient themselves by the sun. However, even under completely covered skies homing pigeons maintain some orienteering ability. It is this ability that seems to be impeded by a magnet carried along on the flight. It appears, therefore, that homing pigeons are able to switch from an astronomical direction-finding technique to a magnetic one when the prevailing conditions require. The presence of a magnetic sense has also been demonstrated rather convincingly in European robins (F. Merkel, W. Wiltschko) and in bees (M. Lindauer, H. Martin).

It will be necessary to address this matter seriously. Unfortunately, even the discovery of a "compass sense" in homing pigeons is not of great help. Such a sense is of no use in determining position—an indispensable factor for finding the way home. The only geophysical phenomenon that has been suggested in this respect (G. Ising) and that in theory would make possible the determination of the geographical latitude is the Coriolis force, a force of inertia (produced by the earth's rotation) to which are exposed all bodies that move in relation to the surface of the Earth. A bird flying east along the equator at a speed of 64 kilometers per hour weighs about 0.2 percent less than a bird flying west at the same speed. If the bird flies north, the Coriolis force will try to deviate it to the right; if it flies south, to the left. As the North or the South Pole is approached, the force gradually fades down to zero. The Coriolis force is responsible for the fact that cyclones move counterclockwise in the northern hemisphere and clockwise in the southern hemisphere. It also has a well known effect on ballistic curves. The Coriolis force has hardly been considered as navigational aid for birds. It is rather improbable that a vigorously moving bird exposed to the wind could take quantitative measures of very weak lateral forces. Besides, the Coriolis force decreases linearly with the speed of the bird and reaches zero when the bird rests. It could therefore not be of any help for direction finding in a cage or before flight.

One other possibility, not yet discussed in the literature, could explain at least a good part of the navigational technique of the homing pigeon: Perhaps homing pigeons and other birds are able to see the stars in the daytime. After all, the brightness of the diurnal sky derives only from the fact that air particles disperse the sunlight and guide it to the eye from all directions, so that the eye is flooded by light impulses. Seen from a high-flying airplane, the sky is black and strewn with stars. By the same token one can, when descending into the shaft of a deep well, see the stars during the day, since lateral light cannot reach the eye. The same is true when one looks through a telescope in daylight. It is often possible to recognize individual stars in the early morning or the late afternoon, when there is still plenty of light. This shows that the amount of scattered light that would have to be screened off in order for the stars to be visible is not very large. The decisive question is whether the bird's eye is equipped with some device that can filter out scattered light. A close look at the ophthalmic structure of birds reveals that all bird eyes have a strange comblike organ that protrudes obliquely from the retina into the interior of the eye. Until now we never knew what to make of this pectiniform organ. Because it throws a shadow onto the retina, it was taken for a biological construction flaw. From the viewpoint of physics, this organ would be suitable for filtering out laterally incident scattered light by letting through light from a very narrow solid angle only. It is easy to confirm this by looking through a comb and then turning it slowly until no more light comes through. If a migratory bird were able to see the stars in the daytime, it could at any given moment determine the geographical latitude, the direction, and—by using its sense of time—the geographical longitude.

Many aspects of animal navigation remain secrets. The navigational principle of whales, which have neither good eyesight nor an appreciable sense of smell (though they have a very good sense of hearing), remains a complete mystery. Neither do we understand the navigation technique of turtles, or that of many fishes; they probably use astronomical points of reference, but it cannot be excluded that weak electrical currents produced by the motion of electrically conducting seawater in the magnetic field of the Earth play some part.

One reason why many secrets of bird navigation have not yet been cleared up is probably the fact that we have no sense of navigation. But before the first Europeans reached the Pacific, the natives of Polynesia had—without compass, sextant, or clock—already explored all the islands between Hawaii, Easter Island, and New Zealand. We have definitive proof that the Polynesians made many long, planned voyages, covering distances up to 2,000 nautical miles. It is entirely conceivable that they used a navigation technique similar to that of some migratory birds.

The art of navigation has fallen into oblivion among the Polynesians, but the method applied by their ancestors can be reconstructed from old travel accounts. Like all good techniques, it is simple. It was based on the fact that the stars always rise in the same spot on the horizon, independent of the season. The Polynesian captains were familiar with a great number of constellations, and they knew at what time of the year and the night each one of the stars would rise. Above all, they remembered which star would rise over the island they were headed for. That star served as a navigational aid, but only as long as it stood low over the horizon. As soon as it rose too high, the course was set by the next star in line on an imaginary star path which the seafarers had imprinted on their memory. During the day, the Polynesians steered by the sun. But they also had recourse to geophysical observations. They studied swells, light reflections of islands on the cloudless sky, or cloud formations which indicated land. Were they using the method of the migratory seabirds?

The light of a rising star does indeed always come from the same direction. Also, the starlight that reaches the surface of the Earth always retains the same angle to the Earth's axis. It is strange that no consideration is given to this interesting fact in the common navigational methods. Is this a key to the navigation of birds? Surely the concept that migratory birds flying by night simply steer toward genetically imprinted constellations on the horizon seems to make more sense than the idea that they twist their heads around to measure the position of the pole star.

How could the above-mentioned idea of the Polynesian navigational method be transformed into a useful and testable hypothesis? Can anything observed in bird flight be traced back to it? For this purpose it is practical to start out with the assumption that certain species of birds had their genetic memory imprinted with the knowledge of which rising stars to use as guides to get from their breeding grounds to a set destination. This navigational mechanism would operate without any problems for many generations, but an ever-increasing navigational error would, quite unnoticeably, be creeping in: With the passing of long periods of time, the stars would no longer rise in the same spots. Their positions would be slowly shifting as a consequence of the precession of the Earth's axis.

Precession is the result of the Earth's rotation and of the gravitational pull of the sun. It operates on a spinning top before it falls over from the Earth's gravity. Precession causes the Earth's axis, which at present points to the North Pole, to rotate in the course of 26,000 years around a cone with the aperture angle of 47°. In 13,000 years the present pole star will have shifted north by 43° and the new pole star will be Vega. Precession moves counter to the west-east rotation of the Earth. This means that the positions of the stars gradually shift as if they were traveling in a circular orbit clockwise above the surface of the Earth. The diameter of this orbit would be about 5,200 kilometers. The present autumn sky will be the spring sky 13,000 years from now, and the stars of central Africa will be shining above Europe.

If migratory birds navigate by the stars, a gradual shift in the location of their winter quarters is unavoidable. Perhaps the routes of migratory birds were in part created by this systematic navigational error, and not only by biological necessity. A more thorough study of these routes

Migrating brown pelicans. Each consecutive bird flaps its wings with a slight delay, so that the wing strokes travel like a wave along the chain.

provides some good arguments for this hypothesis. Many birds do not simply fly to warmer climes, but travel much farther. They often fly beyond the equator and head again for moderate zones. Then there are the numerous land birds that cross the open sea in daring and purposeful flight. How did the first bird that dared fly off over unknown waters find a group of tiny islands across an ocean? The hypothesis that a navigational error caused by the Earth's precession influences the itineraries of birds offers a surprisingly simple explanation: Birds must have attempted again and again to follow their lodestars with blind trust, even if they were lured far out over the life-threatening ocean. Many must have perished in the attempt, but some found a new living space which has by now become their regular destination.

It may even be that the circular precession orbit of the stars is mirrored in the routes of some birds of passage. The golden plover of western Canada flies a loop-shaped path that leads south in the east and north farther to the west. The large shearwater of Tasmania, which circles the Pacific, also flies clockwise. Many North American birds take a more westerly route when they return to the north, and we know of no convincing reason for this choice.

The possibility that a systematic navigational error had a not unimportant part in determining the evolution of flight paths would be a graphic example of how nature can fall victim to its own physiotechnical wealth of imagination.

Nature, Human Technology, and the Future

All the fundamental areas of physics, with the exception of the life-threatening high-temperature and atomic physics, have been opened up and utilized to advantage by living things. Almost all the basic priciples of mechanics, dynamics, thermodynamics, optics, and acoustics had already been in the service of life for millions of years before the human mind learned to understand and master their functions. Mankind must (somewhat belatedly) realize that, to some extent, our technology repeats development which life has already successfully gone through.

In the conquest of the air and the sea and in the mastery over heat and cold, light, and sound, man-made and living structures must take the same physical laws into account. This explains why independent developments in nature and in technology often show similarities. The wings of airplanes resemble those of birds. The latter in turn resemble the flying apparatus of bats, reptiles, flying fish, or some flying seeds. The optical systems of cameras find their biological parallel in the eye structures of animals. The construction principles used in modern building are encountered, often in more developed versions, among the plants. In the microscopic world of biological building materials we find structures that are beginning to prove ideal in the physical-technical sense.

In many areas, human technology is still lagging far behind nature. The ultrasound echo probes used on ships and the new ultrasonic instruments that have been developed for diagnosing cancer cannot stand comparison with the sonar equipment of bats and dolphins. No submarine can compete in flexibility with the deep-diving sperm whale. No technological mechanism can compare to the noiseless, clean, and efficient power of muscles.

Nature devotes great patience to inconspicuous phenomena. Every physicist knows that organic material can destroy the surface tension of water, but who would have thought of using this effect to develop the smooth driving mechanism found in some insects? Who would have been able to exploit, as nature does in trees, the fact that water evaporates more slowly from hairlike tubes?

Not all developments of modern technology can be compared to natural structures. For example, life was never tempted to leave the atmosphere. Neither has evolution ever ventured into high-temperature technology, with its assortment of production processes and combustion engines; but the reaction motor of the bombardier beetle proves that nature could have progressed in this direction had it been of advantage to life.

In places where we would never have deemed it possible, nature offers elegant and often obviously better alternate solutions. Nature, for instance, uses porous elastic layers to secrete a lubricant under pressure and apply it to the gliding surface of mechanical bearings. Nature's optical lenses are not lumenized by a continuously applied film, but by the many submicroscopic, transparent attachments that create a continuous transition from the refractive index of air to that of the lens. In the field of fluid mechanics, the principle of the elastic skin of the dolphin which serves to subdue turbulence has already been tried out successfully in pilot tests. Engineers who are battling noise may find new and important keys to the solution of their problems in the muffling surfaces of owls and nocturnal butterflies. Materials experts ought not to limit themselves to experiments with new synthetics: In its many building materials, nature demonstrates precisely what is required for stability or for elasticity. In the conquest of the deep sea, too, the teachings of nature could be of inestimable help: Perhaps we could learn how to extract air for underwater laboratories from the sea water surrounding them.

The number of interesting prototypes offered by biological systems is very large indeed. But nature does not go about its construction work for the sake of man. We therefore ought to look to nature not for ready solutions, but for inspiration and ideas. Our imaginations are often restricted by training and personal experience. Observation of nature could present us with new dimensions of experience, for in nature we find applied physics free of all inhibitions. Life experiments like a child at play.

Bibliography

Note: Titles of articles in foreign journals are translated; however, titles of foreign books are given in the original language.

Chapter 2

Alexander, R. McNeill. *Animal Mechanics.*. London, 1971.

Cloudsley-Thompson, J. L. The size of animals. *Science* (April 1969): 24.

Collias, N. E., and E. C. Collias. Evolutionary trends in nest building by the weaver birds. In *Proceedings of the Ornithological Conference, Ithaca, 1962.* Baton Rouge: American Ornithological Union.

Eibl-Eibesfeld, J. *Grundriss der vergleichenden Verhaltensforschung.* Munich, 1967.

Elliott, G. F. Variations of the contractile apparatus in smooth and striated muscles: X-ray diffraction studies at rest and in contraction. *J. Gen. Physiol.* 50 (1967): 171.

Frisch, K. von. *Tiere als Baumeister.* Berlin, 1974.

Grzimeks Tierleben, Enzyklopädie des Tierreiches, vols. I–XIII. Munich and Zurich, 1972–1974.

Helmcke, J. G., and W. Krieger. *Diatomeenschalen im elektronenmikroskopischen Bild,* vols. I–III. Weinheim/Bergstrasse, 1961.

Huxley, A. F. Muscle structure and theories of contraction. *Progr. Biophys., Biophys. Chem.* 7 (1957): 255.

Johnson, A. W. *The Birds of Chile.* Buenos Aires, 1965.

McCutchen, C. W. Animal joints and weeping lubrication. *New Scientist* 15 (1962): 412–415.

Nachtigall, W. *Phantasie der Schöpfung.* Hamburg, 1974.

Preston, R. D. The structure of plant polysaccharides. *Endeavour* 23 (1964): 153.

Weis-Fogh, T. (Elasticity.) *J. Mol. Biol.* 3 (1961): 520, 648.

Chapter 3

Alexander, R. McNeill. *Animal Mechanics.* London, 1971.

Bucher, M. Das Bumerang-Werfen. *J. Ethnol.* 48 (1916).

Bainbridge, R. The speed of swimming of fish as related to the size and to the frequency and amplitude of the tail beat. *J. Exp. Biol.* 35 (1958): 109.

Berg, H. C. How bacteria swim. *Sci. Am.* (August 1965).

Denton, E. J. Swimming apparatus of marine animals. *Endeavour* 22 (1963): 3.

Galloway, W. The motion of whales during swimming. *Nature* 116 (1925): 431.

Gawn, R. W. L. Fish propulsion in relation to ship design. *Trans. Inst. Nav. Arch.* 92 (1950): 323.

Gray, J. *How Animals Move.* Harmondsworth/Middlesex, 1959.

Hartman, F. A. *Locomotor Mechanisms of Birds* (Smithsonian Institution publication 4460).

Hertel, H. *Struktur, Form und Bewegung.* Mainz, 1963.

Holst, E. von. *Zur Verhaltenphysiologie bei Tieren und Menschen, Gesammelte Abhandlungen,* vol. II. Munich, 1970.

Kramer, M. The dolphin's secret. *New Scientist* 7 (1960): 118.

Kramer, M. O. Hydrodynamics of the dolphin. *Adv. Hydrosci.* 2 (1965): 111.

Lawson, D. A. Pterosaur from the latest Cretaceous of West Texas: Discovery of the largest flying creature. *Science* 187 (1975): 947.

Lighthill, M. J. How fish swim. *Endeavour* 29 (1970): 77.

Lockley, R. M. *Ocean Wanderers.* Newton Abbot, 1974.

Marchin, K. E. Wave propagation along flagella. *J. Exp. Biol.* 35 (1958): 796.

Nachtigall, W. *Glass Wings.* Munich, 1968.

Nachtigall, W. *Biological Mechanism of Attachment.* New York, 1974.

Pringle, J. W. S. Insect flight. *Cambr. Mongr. Exp. Biol.* 9 (1957).

Prandtl, L. *Strömungslehre.* Braunschweig, 1956.

Rand, A. L. *Ornithology: An Introduction.* New York, 1969.

Rosen, M. W. and N. E. Cornfeld. Fluid friction in fish slimes. *Nature* 234 (1971): 49.

Schleithauer, W. Kolibris—Fliegende Edelsteine. *München o. J.*

Stever, H. G., and J. J. Haggerty. *Flight.* New York, 1965.

Tributsch, H. Nosedive fishing. *Bild der Wissenschaft* (November 1974): 60.

Urry, D., and K. Urry. *Flying Birds.* London, 1969.

Wardle, C. S. Limit of fish swimming speed. *Nature* 225 (1975): 725.

Wigglesworth, V. B. *The Life of Insects.* London, 1968.

Wolters, B. "Jet propulsion" in the plant kingdom: Ejection of seed by the squirting cucumber. *Umschau* (1965): 544.

Chapter 4

Alexander, R. McNeill. Physical aspects of swim bladder function. *Biol. Rev.* 41 (1966): 141.

Cahlander, D. A., J. J. G. McCue, and F. E. Webster. The determination of distance by echo-locating bats. *Nature* 201 (1964): 544.

Graham, R. The silent flight of owls. *J. Roy. Aeronautical Soc.* 38 (1934): 837.

Griffin, D. R. Acoustic orientation in oil-bird (*Steatornis*). *Proc. Nat. Acad. Sci.* 39 (1953): 884.

Griffin, D. R. Comparative studies of the orientation sounds of bats. *Symp. Zool. Soc. Lond.* 7 (1962): 61.

Hertel, H. *Struktur, Form und Bewegung.* Mainz, 1963.

Kellog, W. N. *Porpoises and Sonar.* Chicago, 1961.

Lilly, J. C., and A. M. Miller. Sounds emitted by the bottlenosed dolphin. *Science* (1961): 2.

Schevill, W. E., and B. Lawrence. Auditory response of a bottlenose porpoise, *Tursiops truncatus*, to frequencies above 100 kC. *J. Exp. Zool.* 124 (1953): 147.

Schroeder, M. R. Models of hearing. *Proc. IEEE* 63 (1975): 1332.

Schwartzkopff, J. Acoustic orientation among animals. *Ergebn. Biol.* 25 (1962): 136.

Thorpe, W. H., and D. R. Griffin. Lack of ultrasonic components in flight noise of owls. *Nature* 193 (1962): 594.

Worthington, L. V., and W. E. Schevill. Underwater sounds heard from sperm whales. *Nature* 180 (1957): 191.

Chapter 5

Folk, G. E. *Introduction to Environmental Physiology.* Philadelphia, 1966.

Gärtner, W. Adaptation to cold among mammals. *Umschau* 72 (1972): 219.

Lange, O. L. Lichen—Pioneers of the cold wastelands. *Umschau* 72 (1972): 650.

Chapter 6

Chouteau, J., and J. H. Corriel. Physiology of deep-sea diving. *Endeavour* (June 1971).

Elsner, R. W. Diving mammals. *Science J.* (April 1970).

Chapter 7

Adam, W. Biological light. *Chemie in unserer Zeit* 7 (1973), no. 6: 182.

Bernhard, C. G. Structural and functional adaptations in a visual system. *Endeavour* 26 (1967): 79.

Calvin, M. Solar energy by photosynthesis. *Science* (April 1974).

Denton, D. J., and M. F. Land. Mechanism of reflexion in silvery layers of fish and cephalopods. *Proc. Roy. Soc. Lond. A* 178 (1971): 43.

McElroy, W. D., and H. H. Seliger. Biological luminescence. *Sci. Am.* (December 1962): 76.

Frisch, K. von. *Tanzsprache und Orientierung der Bienen.* Berlin, 1965.

Fogden, M., and P. Fogden. *Animals and Their Colours.* London, 1974.

Gamow, R. J., and J. F. Harris. The infrared receptors of snakes. *Sci. Am.* (May 1973): 94.

Harris, J. F., and R. J. Gamow. Snake infrared receptors: Thermal or photochemical mechanism? *Science* 172 (1971): 1252.

Hudson, R. D., and M. W. Hudson, eds. *Infrared Detectors.* Benchmark Papers in Optics. New York, 1975.

Tiemann, D. L. Nature's toy train, the railroad worm. *National Geographic* (November 1953): 579.

Wehner, R. How do insects navigate? *Umschau* 75 (1975): 653.

Wickingerschiffshalde Museum, Roskilke, Denmark. Model and explanation of the Viking sunstone.

Zahl, P. A. Fishing in the whirlpool of Charybdis. *National Geographic* (November 1953): 579.

Zahl, P. A. Sailing a sea of fire. *National Geographic* (July 1960): 120.

Chapter 8

Gerardin, L. *Natur als Vorbild—Die Entdeckung der Bionik.* Munich, 1968.

Harder, W. Electric fishes. *Umschau* 15 (1965): 467; 16 (1965): 492.

Lissman, H. W. On the function and evolution of electric organs in fish. *J. Exp. Biol.* 35 (1958): 156.

Lissman, H. W. Electric location by fishes. *Sci. Am.* 208 (March 1963): 33.

Lowenstein, W. R. Biological transducers. *Sci. Am.* (August 1980): 2.

Marchin, K. E., and H. W. Lissman. The mode of operation of the electric receptors in *Gymnarchus niloticus. J. Exp. Biol.* 37 (1960): 801.

Wiener, N. *Cybernetics.* Cambridge, Mass., 1948.

Chapter 9

Schildknecht, H., E. Maschwitz, and U. Maschwitz. The explosive chemistry of the bombardier beetle. *J. Nat. Sci.* 23b (1968): 1213.

Schmidt-Nielsen, K. *Animal Physiology.* Englewood Cliffs, N.J.

Scholander, P. F. Tensile water. *Am. Scientist* 60 (1972): 584.

Wigglesworth, V. B. *The Life of Insects.* New York, 1978.

Chapter 10

Emlen, S. T. The stellar-orientation system of a migratory bird. *Sci. Am.* (August 1975).

Frisch, O. von. *Animal Migrations.* New York, 1969.

Keeton, W. T. The mystery of pigeon homing. *Sci. Am.* (December 1974).

Matthews, G. V. T. *Bird Navigation.* Cambridge, 1968.

Orr, R. T. *Animals in Migration.* New York, 1970.

Frey-Wyssling, A. "Right" and "left" in the plant kingdom. *Biologie in unserer Zeit* (1975), no. 5.

Nicolis, G., and J. Portnow. Chemical oscillations. *Chem. Rev.* 73 (1973): 365.

Tributsch, H. Parametric energy conversion: A possible universal approach to bioenergetics in biological structures. *J. Theor. Biol.* 52 (1975): 17.

Index